1795

400

THE

LIGHT

AT THE EDGE

OF

THE UNIVERSE

THE

LIGHT

AT THE EDGE

OF

THE UNIVERSE

DISPATCHES FROM
THE FRONT LINES OF COSMOLOGY

MICHAEL D. LEMONICK

PRINCETON UNIVERSITY PRESS
PRINCETON, NEW JERSEY

Published by Princeton University Press, 41 William Street,
Princeton, New Jersey 08540
Copyright © 1993 by Michael D. Lemonick
All Rights Reserved

Originally published by Villard Books, and reprinted by
permission of Random House, Inc., New York

Library of Congress Cataloging-in-Publication Data
Lemonick, Michael D., 1953–
The light at the edge of the universe : dispatches from
the front lines of cosmology / Michael D. Lemonick.
p.   cm.
Includes index.
ISBN 0-691-00158-8
1. Cosmology.   2. Astrophysics.   3. Astronomers.
I. Title.
QB981.L36   1995
523.1—dc20   94-39324

Princeton University Press books are printed on acid-free
paper and meet the guidelines for permanence and durability
of the Committee on Production Guidelines for Book
Longevity of the Council on Library Resources

First Princeton Paperback printing, 1995

Printed in the United States of America

10 9 8 7 6 5 4 3 2 1

For Eleanor Leah Drutt Lemonick,
who taught me about words,
and Aaron Lemonick,
who taught me about the universe

CONTENTS

Human beings have been gazing at the night sky, wondering about the lights they saw there, for many thousands of years. Ancient skywatchers observed the slow, curving march of stars across the sky from dusk to dawn. They noted even slower changes from season to season. And they became aware of the planets, wandering along with the Sun and Moon through the fixed background of stars.

Once language originated, people started to tell stories about the heavens—stories that attempted to explain what these mysterious objects were, how they came to populate the sky, and why they formed one particular set of patterns and not another. The stories are cosmologies—descriptions of the universe, its origins, and its history—and which story people accept has always depended largely upon when and in which culture they grew up.

When I was growing up in the 1960s, I learned the cosmology of modern Western science based upon observations that began with Arab, Greek, and European astronomers, and informed by the laws of physics described by Newton, Einstein, Bohr, Pauli, and others. According to this story, the stars, like the Sun, are huge balls of mostly hydrogen gas, cooking with the fires of nuclear fusion; the galaxies are enormous clumps of stars, and just about every galaxy is rushing away from every other in the aftermath of the Big Bang, the event that started the history of the universe.

Big Bang cosmology was persuasive enough to convince most physicists and astrophysicists that it was correct, but it was based upon remarkably little evidence. The evidence was probably greatest for the Big Bang itself. It was already clear in the 1920s that every galaxy astronomers observed was moving further away from every other galaxy, as though they had all started at a single point and been flung outward. In 1964, observers found the almost imperceptible faint glow of light pouring in from every part of the cosmos. It was presumably left over from a time billions of years ago, when the entire universe was young, compact, and white-hot. Nothing else but a primordial Big Bang could comfortably account for either the light or the expanding universe.

But as for the details—the processes by which a hot, dense universe expands and evolves into a mostly dark place dotted with the light of galaxies—there was not much cosmologists could say. They had many theories, but very few observations of the universe beyond a very small neighborhood centered on Earth's home, the Milky Way. It was not for lack of trying, but the telescopes and other astronomical tools at their disposal were neither numerous nor powerful enough to supply much information. What information they did have, moreover, was confusing. During the 1970s, for example, most astronomers finally accepted what a few had been saying for decades, that there was some sort of invisible "dark matter" surrounding many galaxies, and the dark matter outweighed the visible stars by a considerable margin.

It was not until the mid-1980s that astrophysicists, armed with more telescopes, Earth-orbiting observatories, sensitive electronic light detectors, and powerful supercomputers, finally began to attain the data they had so desperately wanted. What they saw was quite unexpected. It was common knowledge that galaxies usually did not travel alone; scores of these immense clouds of stars would frequently be gathered into clusters of galaxies. But the sky surveys of the early 1980s showed that galaxies were clustered on an unsuspectedly huge scale, into clusters upon clusters of galaxies that no theory had ever predicted.

Somehow, astronomers had to reconcile this unexpected level of cosmic structure with the dark-matter problem, since dark matter exerts gravity and gravity surely must have something to do with how the universe is organized. They also had to explain why the light from the early universe was uniform in all directions; if the universe is highly structured today, it must have been somewhat structured in the past. The light—no longer visible, but now in the form of microwaves—should show some structure as well. Yet it showed none. Again, astronomers had plenty of theories, but none of them seemed to account for the data.

Such was the state of cosmology when I began work on this book in 1991. I had been covering science for *Science Digest, Discover,* and *Time* magazines for most of the decade, and had watched as each new discovery of the universe seemed only to deepen the mystery. I wanted to understand exactly what astrophysicists knew, what they suspected, and how they were going about settling these disturbing questions. It was clear from the outset that I would not be able to visit every observatory or speak with every cosmologist; it was also clear that by speaking to a well-selected group of men and women, theorists and observers alike, at several institutions, by visiting several observatories, and by attending as many conferences, colloquia, and informal gatherings of astronomers as I was able, I could reasonably expect to understand the existing observations, the competing theories, and the major unanswered questions of modern cosmology.

It certainly did not hurt that I lived in Princeton, New Jersey, where there are probably as many astrophysicists within a twenty-mile radius as there are anywhere in the world—and where dozens from other parts of the world pass through in a given academic year. But I also traveled to Berkeley and Santa Cruz, California and Cambridge, Massachusetts, and talked to astrophysicists from many other places by telephone. I visited optical and radio observatories in Chile, Arizona, New Mexico, and West Virginia.

Most important of all, perhaps, was the time I was in Washington, D.C. in April of 1992 for the meeting of the American Physical Society. It was there that astrophysicist George Smoot presented a result that cosmologists had been anticipating for two decades. The Cosmic Background Explorer Satellite, Smoot told an excited audience of physicists and astronomers, had at long last detected hints of cosmic structure in the microwave light of the early universe. "If you're religious," he said later, "it's like seeing God." Modern cosmology was saved by what Stephen Hawking called "the scientific discovery of the century, if not all time."

Or was it? Now, two and a half years later, that is not quite so clear. As cosmologists have discovered more than once in the long history of telling stories about the universe, anyone who claims to have found the ultimate answer is almost certainly wrong. Cosmology is still full of questions, and cosmologists will have mysteries to wrestle with for years. As the hardware of astrophysics and the theoretical cleverness of astrophysicists continue to improve, the next decade should prove even more exciting than the last. It will be fascinating to watch.

Princeton, New Jersey
August, 1994

ACKNOWLEDGMENTS

This book could never have been written without generous help from friends, family, colleagues, and people who were, before the project began, complete strangers. I am indebted first of all to the astronomers, physicists, and technicians who gave me hours of their valuable time, explaining their science and their craft over and over until I finally got it; most of their names and their words appear in these pages. My visits to observatories, conferences and workshops would have been impossible without assistance from staff members at the National Optical Astronomy Observatories, the National Radio Astronomy Observatory, the Institute for Advanced Study, Lick Observatory, Whipple Observatory, the Harvard-Smithsonian Center for Astrophysics, the Center for Particle Astrophysics, AT&T Bell Laboratories, and the departments of physics and astrophysics at Princeton University.

Much credit goes to my friend Andrew Revkin, who encouraged me to leap into the abyss; to Marianne Sussman, whose exaggerated confidence in me led to a lunch with my editor, Douglas Stumpf; to Doug, whose unflappable temperament, sharp eye, keen ear, and acute literary judgment made the editing process not just painless but actively enjoyable; to Michael Carlisle of the William Morris Agency, whose steady hand steered me through the unfamiliar and perilous seas of book publishing; to Marcia Bartusiak, Richard Preston, and Paul Hoffman, whose encouragement was unfailing; to Andrea Dorfman, Philip Muck, Donna Biddulph, Brad Hill, and Mark Murphy, whose thoughtful comments on the manuscript provided reassurance at crucial times; to John Bahcall, of the Institute for Advanced Study, who graciously allowed me to sit in on the legendary Tuesday Lunch; to Philip Schewe, of the American Institute of Physics and Steve Maran, of the American Astronomical Society, for invaluable advice and logistical support; and to Donna Laurie, for saving me a seat.

I also thank Tony Tyson, Ed Turner, Jackie Hewitt, David Wilkinson, George Blumenthal, John Huchra, and Jerry Ostriker for reading parts of the manuscript and David Weinberg for reading the entire thing. Their care and thoughtfulness resulted in the correction of a significant number of mistakes; whatever errors made it into print are entirely my own.

Henry Muller, Dick Duncan, Ron Kriss, Walter Isaacson, Claudia Wallis, and Charles Alexander were extraordinarily considerate in allowing me the freedom to learn more about astrophysics than would ever fit into a *Time* story. And I am not sure how it would all have turned out if Nancy Manning, Tom Baker, Pat Howe, John McPhee, Ken Goldstein, Leon Jaroff, and Bill Wilson had not all, in different ways, shared their wisdom and experience with me along the way.

I thank my parents; my brother, David Lemonick; and my stepson, Ben Hohmuth for their constant support. And finally, I owe the greatest debt of all to Hannah, who has broadened my horizons considerably and is the source of my greatest joy, and Eileen, whose talent produced many of the photographs that I include with pride in this book, but whose love and support through everything would have been more than enough.

THE

LIGHT

AT THE EDGE

OF

THE UNIVERSE

CHAPTER ONE

INTRODUCTION

J udging from the appearance of the nighttime sky, Mount Hopkins, Arizona, could be on a different planet from my home in Princeton, New Jersey. Back East, the stars are sparse and dim, mostly lost in haze, air pollution, and the reflected glare of city and suburban lights. Only a few constellations are visible even on the darkest nights. You only notice them if you make an effort. The bright planets—Venus, Jupiter, and sometimes Mars—are easily confused at a casual glance with the steady parade of airliners and small planes that cross overhead. If the ancient Greeks and Arabs, whose observations underlie most of Western astronomy, had tried to work under these conditions, they would probably have given up.

At Mount Hopkins, by contrast, you can't escape the stars. The heavens are carpeted with them, and bisected by the even brighter band of the Milky Way, broad and diffuse but unmistakable. The glow of Tucson, thirty-five miles to the north, of the twin cities of Nogales, Arizona, and Nogales, Mexico, twenty-five miles to the south, and of Interstate 19, which connects the two urban centers, does nothing to diminish them. The stars are so numerous and so bright that they have almost a physical presence. It would have been a little bit unnerving to be alone on the mountain during a visit I made in the middle of February 1991, with what seemed to be a heavy curtain of light pressing down from overhead.

Fortunately, I had a guide. His name is John Huchra, and he is one of the world's hardest-working and most talented observational astronomers. This is his own opinion and the opinion of most of his colleagues. At the time of my visit, Huchra was an administrator as well as a scientist, serving a term as associate director for optical and infrared astronomy at the Harvard-Smithsonian Center for Astrophysics—the CfA for short. *Serving* is the appropriate verb: Huchra's own abbreviated description of the job was "jail." Normally, Huchra spends about 130 nights a year at observatories around the world, but his new position had limited him to only ninety or so, still an enormous number for most astronomers, but a starvation diet for Huchra. He knew the administrative work was important, and that having someone with his drive and working knowledge of observational astronomy in the position would be good for the other observers at the Smithsonian. But like any veteran inmate, he always knew to the day, without bothering to consult a calendar, exactly how much time he had left on his sentence.

On the first night of my visit, the stars began to pop out of the darkening sky almost as soon as the sun went down, and as full darkness came on they flooded the sky. Jupiter glowed like an electric arc to the east; Venus, even brighter, had set an hour earlier. Mars glittered overhead, and to the south was Sirius, the brightest star in the sky, but a poor second to Jupiter. We went inside the observatory building for a minute: Huchra's observing run would not start until two nights later, but the next evening he would be helping a postdoctoral fellow from the center use the telescope for a thesis project, so he had to spend some time this night on the computer, reading his electronic mail from Cambridge. "It used to be," he said, "that when I came observing, I could escape the real world. Now the eighty-seven people who are after me can still get to me. This is the wonder of E-mail." He called up one message after another, sighing every once in a while before tapping out a reply. Then, after a long silence, "Oh, no." It sounded as though something was terribly wrong. "My bowling team has slipped to third place."

Reading the messages would take him a while, so Huchra set me up with a pair of binoculars and went out to show me some of the celestial sights. I asked him to find me the great nebula in Andromeda, and, after a few minutes of triangulating from various guide stars, he pointed to a spot to the left of Cassiopeia.

"Right there," he said, and for the first time (other than in photographs) I saw the only full-fledged galaxy besides our own that is visible to the naked eye. Although it contains one hundred billion stars, it looked in the Arizona sky like nothing more than a fuzzy patch of light, much fainter than the Milky Way, a long and narrow oval stretching about four times the width of the full Moon. Until the early 1920s, astronomers did not even know for sure that Andromeda was a galaxy; many believed it was a cloud of glowing gas within the boundaries of the Milky Way. But it is now known to be a star system in its own right, an island of suns lying about two million light-years away.

A light-year is a measure of distance, and not, as one might guess, of time. It is the distance light travels, going at more than 186,000 miles per second, in one year. It equals nearly 6 trillion miles. The term is rarely used by astronomers. They prefer parsecs, shorthand for "parallax second," which is the distance away a star has to be to shift its apparent position on the sky by a second of an arc—a thirty-six hundredth of a degree—as Earth moves from one side of the sun to the opposite side over six months. Parsec is preferred because parallax is the way distances to nearby stars are measured.

Although the Andromeda galaxy is tens of thousands of times farther away than the individual stars of the Milky Way, there was no obvious way for me to tell, even in the clear skies of Arizona.

The sky has no depth clues, and the most experienced astronomer sees it just as the ancient Greeks or Egyptians or Chinese did, and just as a child does as well: like a black bowl inverted overhead, its inner surface dotted with light. It would be possible to see the third dimension if stars came in one brightness only, like hundred-watt light bulbs do. If that were true, then a star's apparent brightness in the sky would indicate its distance; the closest ones would look brightest, and farther stars would seem dimmer. But in the real universe, a bright star might be average and nearby or brilliant and far away; a dim star may be intrinsically feeble or extraordinarily distant.

Yet for astronomers trying to understand how the universe is put together, and how it has evolved, it is essential to understand how it looks in three dimensions. Imagine a demographer trying to chart the spread of population across the United States over the past century. The most elementary piece of information, the starting point, for such a study would be the distribution of population today—how many people there are and where they live. If you don't know where they are today, you can't very well figure out how they got there.

Astronomers are faced with the same kind of problem. Until they trained telescopes on the Milky Way in the 1600s, they had no idea that its diffuse glow came from millions of stars, and until they found ways to measure the distances to the stars in the 1800s, they had no idea that the Sun and every other star in the sky is part of a single unit called the Milky Way galaxy. It was only by extending that third dimension further, measuring the distance all the way to Andromeda and other, fainter galaxies that Edwin Hubble proved, early in the present century, that the Milky Way is only one of many galaxies in the universe. Charting the directions and distances to other galaxies, Hubble became the first cosmic cartographer, the first to tackle the problem of constructing a modern map of the universe. Making careful measurements with the hundred-inch Hooker telescope on Mount Wilson, on what was then the outskirts of Los Angeles, he measured the distances to about twenty nearby galaxies. His work—he was also the one who discovered that the universe is expanding—made possible the science of cosmology, the study of the universe as a whole.

That was about seventy years ago. Since then, cosmologists have been refining Hubble's measurements and adding many of their own, making photographic surveys of the deep sky and assembling them into atlases that show millions of galaxies. The atlases, like the sky you see with your naked eye, are two-dimensional, though; they show where on the sky the galaxies lie, but not how far away they are. Their distance, the third dimension necessary to making a realistic map of the universe, is missing. The reason is that it is relatively easy to take a picture of the sky and capture a million galaxies at once, but very tedious and difficult to measure the distances to galaxies one by one. "In the early 1970s, when I started thinking about this problem," Huchra had told me when we met in his office in Cambridge a few months earlier, "there was essentially zilch data on galaxy distances. The most complete sample had 273 galaxies in it—not an overwhelming number."

By the end of the 1970s, thanks largely to Huchra and another astronomer named Marc Davis, the sample had expanded tenfold, and there were hints that the universe was more complicated than anyone had expected. Astronomers had always assumed that matter was, by and large, distributed evenly through space. On small scales, of course, that is not true. Matter in the Solar System comes in lumps—the planets and the Sun—with mostly empty space in between. The stars of the Milky Way are widely spaced lumps of matter, too, and the galaxies in turn are lumps in intergalactic space. There are clusters of galaxies as well, where galaxies are crowded together to form even larger lumps. But on really large scales, in collections of thousands or millions of galaxies, it was presumed that everything would even out, just as individual grains of sand merge into a smooth stretch of beach when they are viewed from a distance.

What Huchra and Davis started to see in the late seventies were hints that even on enormous scales the universe was somewhat uneven, with large areas that had more galaxies than average and large areas that had fewer. Other astronomers were seeing similar effects. Most notably, the husband-wife team of Riccardo Giovanelli and Martha Haynes had used radio telescopes to map an enormous chain of galaxies, which they called the Perseus-Pisces Supercluster, that stretched across the southern sky. Two thousand galaxies is a bigger sample than two hundred, but it is still not large enough for astronomers to say anything meaningful about the large-scale structure of the universe. So in the mid-1980s, Huchra started in on a much bigger survey, now working with a Harvard astrophysicist named Margaret Geller. And by 1986, with ten thousand galaxies in their catalog, Geller and Huchra and a graduate student, Valerie de Lapparent, were ready to present the first detailed map of a significant chunk of the universe. In January, Geller showed up at the winter meeting of the American Astronomical Society in Houston and gave a talk titled "A Slice of the Universe."

It made headlines—and not only in professional science journals. It appeared on the front page of *The New York Times* and in the pages of *Time* magazine, and Geller was whisked off to New York to appear on the *Today* show. The universe revealed in the map was not the one astronomers had expected. The slice of the universe charted by Huchra, Geller, and de Lapparent, tens of millions of light-years deep, was the biggest chunk of cosmos ever mapped in detail, and it was more structured than anyone expected—even though they had been prepared by the earlier surveys. The galaxies in the map were huddled together to form narrow filaments. In between the filaments, there was hardly anything at all—just enormous voids of empty space. The voids were more or less round, as though the astronomers' slice had cut across a froth of soap bubbles. And the walls of these bubbles were sharply defined. No theory had predicted the existence of structures this big, and no theorist could suggest any mechanism to explain how they had come about.

Huchra, Geller, and de Lapparent's bubbly slice of the universe was not the only jarring news that astronomers had to deal with that year. Sandra Faber, an astronomer from the University of California, Santa Cruz, gave a talk at a later conference in Hawaii, reporting on the observations of a team of seven astronomers (including herself) that subsequently became known as the "Seven Samurai." They, too, had been surveying galaxies, but they were interested more in motions than in distances; they wanted to see how galaxies in the neighborhood of the Milky Way are moving through space. Overall, the universe is expanding, and any two galaxies you choose will, on average, be moving apart as it does. Locally, though, gravitational attraction between galaxies can overcome the general expansion, creating "peculiar motions" that are superimposed on the expanding universe. Peculiar motions can make galaxies separate more slowly than they otherwise would, or even overcome the expansion altogether—the Milky Way and the Andromeda galaxy, for example, will have a close encounter, or even collide, in ten billion years or so. Clusters of galaxies, more massive than single galaxies, exert stronger gravity and create larger peculiar motions. The Milky Way and Andromeda, besides approaching each other, are together being pulled in the direction of the great cluster of galaxies in the constellation Virgo.

What the Seven Samurai did was to look at the peculiar motions of galaxies in an enormous volume of space, hundreds of millions of light-years across. They found that hundreds of galaxies, including the Milky Way, are all falling through space toward a point marked on the sky by the constellations Hydra and Centaurus. Astronomers already knew that the Milky Way was moving in that direction, another peculiar motion superimposed on the general expansion, and they thought they knew the reason: there is a large cluster of galaxies in Hydra-Centaurus, and its gravity was undoubtedly pulling the Milky Way toward it. But the Samurai discovered that Hydra-Centaurus was moving, too, and in the same direction. Something was pulling on it—something farther away, and much bigger. Alan Dressler, another one of the Samurai called this enormous chunk of matter, presumably a giant knot of galaxies, the Great Attractor. Like the bubbles and voids in the Harvard-Smithsonian survey, such an enormous structure was unexpected by most astronomers. It came as a shock.

Either one of these discoveries by itself might have been dismissed as a quirk, a kind of joke being played by Nature. If all astronomers had to go on was the Harvard-Smithsonian survey, they could have argued (and some did at first) that the universe really is smooth on large scales, and that Huchra, Geller, and de Lapparent had just happened on an area that was uncharacteristically full of structure. If all they had was the Great Attractor, they could say the same. With both phenomena, though, the evidence was compelling that the universe is structured on scales larger than anyone had previously imagined.

If astronomers' presumption of large-scale cosmic smoothness had been based purely on imagination and prejudice, they could simply have adjusted their worldview, just as an earlier generation of astronomers had done when they were forced to accept the Milky Way as just an average galaxy among many, rather than the only one in the universe. But there already existed direct evidence that the universe really is smooth on very large scales. That evidence is known as the cosmic microwave background radiation, and it is the echo of the Big Bang itself, still reverberating through the universe. It is the oldest light in the universe—for microwaves, like any form of electromagnetic radiation, including X-rays, infrared light, ultraviolet light, and ordinary visible light, are all basically the same stuff at different energy levels. For the first few hundred thousand years of its life, the universe was filled with both electromagnetic radiation and a hot, thick soup of elementary particles and atomic nuclei. The radiation couldn't penetrate the soup. But as the universe expanded and temperatures dropped to about five thousand degrees Kelvin (coincidentally, about the same temperature as the surface of the Sun), the soup thinned out, and the radiation burst free. With nothing appreciable to stop it over the billions of years between then and now, this light is still with us, weakened and cooled, but still detectable.

The discovery in 1965 of that feeble but pervasive radiation, in the form of microwaves (the same kind emitted by microwave ovens and radar transmitters), streaming in from all directions, turned the Big Bang at a stroke from a plausible theory into a widely accepted model for how the cosmos began. According to the model, the universe started in a condition of infinite heat and density (a condition beyond human imagination, and also beyond the descriptive boundaries of physics) and proceeded to expand and cool, the temperature and pressure dropping all the while. After perhaps .000000000000000000000000000000000000000001 second that science can't yet deal with, the known laws of physics began to apply. They predict, among other things, that the microwave background should exist and that the universe should be expanding—both of which are observed, and neither of which is convincingly predicted by any other model. It's a "model" or a "theory" rather than the absolute truth, because no one can say for sure whether another idea won't work better, predicting the same observed phenomena that the Big Bang model does, but explaining other still-mysterious facts about the universe as well. Astronomers are currently proceeding on the assumption that the Big Bang is basically right, but most of them acknowledge that it could be modified or, conceivably, even overthrown someday.

After the euphoria that came with discovering the cosmic microwave background had worn off, physicists and astronomers realized that it provided a way to probe the structure of the early universe. If there were any sort of structure in the universe at the time the cosmic background radiation broke free—any pattern of high and low density that departed from perfect smoothness—the radiation should still bear the imprint of that structure. Areas of higher density, with more matter in them, should have trapped correspondingly more radiation, and areas of lower density should have trapped less. Traveling unimpeded through space since then, the pattern of warmer and cooler patches of microwaves, perfectly reflecting the pattern of higher and lower densities, should still be visible today.

They aren't—or weren't when the Harvard-Smithsonian and Seven Samurai astronomers presented their surprising results. That didn't mean the density fluctuations didn't exist—only that they were too small to be measured, that the universe was too smooth early on for any structure at all to be visible. Of course there must be structure at some level; that, too, had been presumed for more than a decade. Otherwise there wouldn't even be galaxies, let alone clusters of galaxies. This was the foundation of what some astronomers called the standard model of the universe. It is made of two basic elements: a Big Bang and gravity. First came the Big Bang, an explosion of unimaginable violence and power, in which the entire cosmos was born. Then, as the pure energy of the explosion expanded and cooled, and condensed from energy into matter, tiny regions of slightly higher density than average began to grow. Their gravity pulled in the surrounding matter, and they grew steadily larger until they formed the concentrations of mass that eventually became galaxies.

In this simplified form, the standard model is reasonable. It almost had to have happened this way. The microwave background says there was a Big Bang. (So does the expanding universe, which, run backward in the imagination, points to a moment when the cosmos must have been in a state of tremendously high density and temperature.) And gravity is a force that naturally builds big objects out of little ones—planets, for example, out of the dust and gas that constituted the young Solar System. Gravity needs something to work with, though, some molehill out of which to build mountains. If the early universe had been truly and perfectly smooth, with no region even a hair denser than any other, then the modern universe would be made of lone atoms, each one equally spaced from its neighbor. The one element missing from the standard model was some evidence of these density fluctuations, some imprint on the cosmic microwave background; the question was, how closely would astronomers have to look before they would see them.

By 1986, the theorists who were trying to make sense out of the raw astronomical data supplied by people like Huchra were already aware that they had some explaining to do. Considering the amount of structure known to exist in the universe—the clusters and superclusters of galaxies observed decades before the Slice of the Universe—some pattern should already have been detected in the cosmic microwave background. The fact that it wasn't was disconcerting.

But theorists make their living wriggling out of tight intellectual spaces; when a theoretical model doesn't quite work, they can make a few adjustments. The standard model wasn't working. The adjustment, well in place by the time the Harvard-Smithsonian Slice of the Universe survey was published, has come to be known as the cold dark matter (CDM) model. The CDM theory originally postulated by P.J.E. Peebles, at Princeton, and since elaborated by many others, fixed the problems with the standard model (and a couple of other problems in astrophysics as well) with a number of remarkable assertions about the universe, not one of which has ever been proven. First, CDM claims that the stars and galaxies astronomers can see make up approximately 1 percent of the mass of the universe. The rest is utterly invisible. Second, this dark matter is made of a type of elementary particle that has never been detected in a laboratory or anywhere else. The particles are "cold"—which is to say, in the jargon of physics, that their characteristic speed as they whip through the universe is much slower than the speed of light. Third, the universe went through a period of incomprehensibly rapid expansion before it was far into its first second of life, a sort of turbocharging of the already explosive expansion of the Big Bang. Fourth, the galaxies formed not everywhere, but only in the most densely packed regions of dark matter. The galaxies are like a frosting of bright snow on the very highest peaks of the dark mountains of particles that fill the universe.

These four statements were unproven, but not preposterous; there were theoretical arguments, at least, to support all of them. They appeared to reconcile the apparent contradiction between a lumpy universe and a smooth microwave background because cold dark matter particles, if they really existed, barely interacted with electromagnetic radiation at all. Unlike ordinary matter, which would leave an indelible imprint on the cosmic background radiation, cold dark matter could clump under gravity to form the seeds of modern structure without marking the cosmic microwave background —or at least not enough for the marks to be visible in any of the detection experiments that had been done up until then.

The new theory—the standard model with cold dark matter added to it—worked so well that it was quickly adopted by most astronomical theorists. It became, in effect, the new standard model. "When you're finally on to a theory that's right," a Princeton astronomer named Ed Turner told me one day soon after my visit to Mount Hopkins, "there is a strong sense that everything is coming together. Every new thing seems to support the theory. And when you see that, you get a real sense that you're on the right track." With cold dark matter, it looked as though astronomers finally understood the mystery of why the universe looks the way it does.

Unfortunately, the universe refused to cooperate. Under the rules of cold dark matter the cosmos should end up looking a certain way; there should be a hierarchy of structure, with galaxies huddled together into clusters of galaxies, and the clusters gathered into bigger clusters. The amount of clustering on these different scales was prescribed by the new model. But when observers like Huchra and Faber looked deep into space, they found giant bubbles lurking in slices of the universe, and they found the Great Attractor. In a universe dominated by CDM, both of these objects are too big to exist comfortably. That didn't mean they were impossible, but it did mean that many more of them, or bigger objects, would mean trouble.

Things have only gotten worse for CDM since then. Alumni of the Seven Samurai, among others, have decided that the Great Attractor itself is falling into an even bigger and more distant chunk of matter known as the Shapley concentration. When Geller and Huchra sliced into the universe more deeply and at wider angles, their new map confirmed that the bubbles in the original survey were real and widespread—not just a fluke. And right in the center of the map, spreading across nearly a third of the sky, was something that was quickly labeled the Great Wall. It was a sheet of galaxies five hundred million light-years long, two hundred million wide and about fifteen million thick. That would make it by far the biggest organized structure ever seen in the universe, and it might be even bigger than it looks: the Great Wall runs right to the edge of the map, so it could easily extend far beyond.

Other astronomers, meanwhile, were making other sorts of distance surveys, and they were disturbing, too. Two groups, one in England and one in the United States, created maps of galaxies that went far deeper into the universe than Geller and Huchra had, though they covered a much smaller expanse of sky. These long, thin "pencil beam" surveys showed what seemed to be not just one but a whole series of great walls of galaxies, interrupting the relative emptiness of intergalactic space every 360 million light-years or so.

Trouble began showing up from the past as well. Observers who specialize in quasars, points of light so bright they can be seen across half the universe, have been finding them farther and farther away. There are now ten quasars known to be so distant that the light reaching Earth has been traveling since the universe was only about 15 percent of its present age. The quasars are sending information from farther back in the universe's history than anything, other than the cosmic microwave background. But because quasars are thought to be the bright cores of young galaxies, their existence means the universe was making galaxies long before the cold dark matter theory says it should have been able to.

Yet another challenge to the cold dark matter model came from a sort of statistical characterization of clumpiness that emerges when you analyze the positions of galaxies and clusters of galaxies over wide areas of the sky. These are two-dimensional positions, without depth, but for this sort of analysis it doesn't matter. The idea is to see whether galaxies (or clusters) are closer together, on average, than chance would predict, and if so, how much closer. Several different groups of astronomers, looking at different samples, have all decided that the universe is clumpier than CDM predicts it should be. Significantly, one of the groups includes Carlos Frenk and George Efstathiou, half of the original "Gang of Four," who were among the earliest and most enthusiastic proponents of the CDM theory (astrophysicists, it should be clear by now, have a special affinity for catchy names both for themselves and for the cosmic objects they look for).

Finally, CDM has been squeezed to the point of discomfort by new, ever more precise measurements of the cosmic microwave background radiation. Microwave detectors on the ground at the South Pole, on board stratospheric balloons and on the Cosmic Background Explorer (COBE), which went into orbit around Earth in 1989 have probed deeper than ever before into the microwave background. By late 1991, the collective data from these devices led to the opinion that nowhere in the sky does the microwave background vary from the average by more than a few parts in ten thousand, a number permitted by the equations of CDM, but just barely.

An outside observer might reasonably have concluded that after only half a decade, the theory was in trouble, and that astronomers had better start shopping for another one. And in fact, by the time John Huchra went to Mount Hopkins for his observing run in February 1991, *Nature,* the prestigious British scientific journal, had proclaimed that COLD DARK MATTER MAKES AN EXIT, while the

American Astronomical Society had hastily called a press conference at its meeting the month before so scientists could brief journalists on the question of how much trouble CDM was really in.

Yet to abandon an idea that just a few years earlier had convinced astronomers that, as Ed Turner said, "everything is coming together," seemed to many astronomers like stepping off into an abyss. "You can't get around the fact that there's more structure on large scales than CDM allows," admitted Joel Primack, a particle physicist, who helped hone CDM into its most persuasive form. "We've got to be honest; if CDM is wrong, we have to admit it. But you have to understand that no idea as beautiful as this has ever been wrong." Beyond that, there simply was no good alternative. Other theories had been discarded before CDM appeared ("hot dark matter" was one) because they were even worse at explaining the structure of the modern universe. Still others (involving cosmic strings, cosmological constants and textures, for example) were too farfetched; latching on to them might be a last resort, but only that.

The field of astrophysics had, in short, entered a period of crisis, a situation where, as Jeremiah Ostriker, a respected theorist and the chairman of Princeton University's Department of Astrophysical Sciences told me at the time: "No existing theory can explain the structure of the universe. There is some missing ingredient, some crucial fact that we haven't yet uncovered, and that's becoming more apparent all the time."

The chance that a theory is about to be demolished is galvanizing. Just as the mass extinctions of species in Earth's distant past have led to a burst of experimentation by Nature, and a rush of evolutionary innovation, so has the cosmological crisis spurred intense creative effort on the part of theorists. Some, like Peebles, who came up with the cold dark matter theory in the first place and long since abandoned it, were happy to play around with other ideas. Others—Michael Turner, of the University of Chicago, for instance—were still sure that CDM would, with a little fine tuning, survive nicely. But virtually all the theorists in astronomy, whether they were scrambling to come up with new, more exotic theories of the cosmos or scrambling to adjust CDM to the new reality imposed on them by observations, began attacking the equations of matter and energy with an intensity not seen in a decade, and conjuring up ever-more complicated imaginary universes in their computers to test their theories.

At the same time, such observers as Huchra were returning to the mountaintops, urged on by the theorists but motivated equally by their own need to know what the universe is really like, not in computers or equations but in the reality of deep space. For the observers knew better than anyone that for all the elaborate models and all the impressive technology astronomers have constructed, astronomers know next to nothing about the universe. They don't know what it looks like. ("The fraction of the universe we've mapped so far is comparable to the fraction of the Earth represented by Rhode Island," Huchra's colleague Margaret Geller is fond of saying.) They don't know what it's made of. They don't know how big it is, or how old. They are still asking the same questions, albeit in more sophisticated ways, that ancient philosophers asked.

Theories are easy to invent without the inconvenience of observed fact, and are also largely useless. So the observers try to answer these most basic questions about the universe by heading for the mountains of Arizona, Hawaii, and the Chilean Andes, the hills of West Virginia and Puerto Rico, and the high, dry desert of New Mexico to peer at the universe. There they hope to learn whether there are more Great Walls and Great Attractors, or structures that are even bigger, and whether there is, finally, some pattern in the microwave background. They are trying to look back to the era when galaxies first began to emerge to find out what processes were responsible.

When the theorists lift up their heads from their calculations and simulations, and the observers return from the field, they meet to discuss, or argue, their findings at astronomical meetings and on the colloquium circuit, the informal weekly lectures that take place every week on every campus where the density of astronomers is high enough. Long before an astronomer publishes a scholarly paper in the *Astrophysical Journal* or *Physical Review Letters* or *Nature*—the signal for *The New York Times* and *Time* and the television news to let the public know about it—he or she has usually stood up in front of colleagues, presented the latest result, and faced intensive questioning by the toughest audiences conceivable. It is in these arenas—the observatories, the theorists' offices, and the meetings and colloquia where the two branches of astronomy interact, that the cosmological revolution is taking place, and where the crisis in cosmology will be resolved.

CHAPTER TWO

WHO NEEDS DARK MATTER?

O f all the claims made by the cold dark matter model, the least controversial by far is its assertion that most of the universe is invisible. At least 90 percent of the cosmos, by weight, and perhaps as much as 99 percent, has escaped detection. No one has ever seen it. It gives off no radiation whatever: no infrared, visible, or ultraviolet light, no radio waves, no X-rays, no nothing—or, at least, so close to nothing that it has never been found, despite careful searches with every sort of telescope known to astronomy. Yet virtually every astronomer is convinced that the dark matter exists. Dark it may be, but it is matter, and that means that it exerts gravity on other types of matter, including the stars and galaxies and clouds of gas that we once thought was all there was to the universe. Like H. G. Wells's Invisible Man, the dark matter's presence can be inferred by the ways it knocks visible things around. If a star or a galaxy appears to be responding to a gravitational tug, there must be something doing the tugging, whether or not the something can be seen. (There is actually another possible explanation: maybe the laws of gravity behave in ways we don't understand. A very few physicists think this might be the case, but not enough to make the idea convincing.)

The idea that there is dark matter in the universe is not particularly new. The first observer to claim that he had found some was Fritz Zwicky, a Swiss astronomer on the faculty at Caltech. Back in the early 1930s, Zwicky had pointed a telescope at the Coma cluster of galaxies, which is now known to be just part of the Great Wall in the northern sky, and measured how fast the galaxies in it were moving in relation to each other. These local motions would reveal how much mass the Coma cluster had in it. The faster the galaxies were swirling around, the more mass the cluster must contain.

What Zwicky found was that the galaxies were moving too fast. Judging from the amount of visible matter in Coma, they should long since have flung themselves out into intergalactic space. If the cluster really was a cluster, and clearly it was, then something must be reining in the galaxies, something that had plenty of mass but did not shine with any light that Zwicky could see. There was, he decided, about ten times as much invisible matter, type unknown, as there was visible matter in the Coma cluster.

It was an astonishing result, but it changed astronomical history only in retrospect. The measurements were terribly difficult to make with the telescopes and photographic plates of Zwicky's era, and so were open to question. Moreover, Zwicky had a dual reputation. On one hand, he was considered ingenious and capable of piercing insights into a broad range of astronomical problems. On the other, he was more than a little eccentric. Worst of all, he was known to attack colleagues for stealing his ideas, so not many astronomers were willing to pursue a line of investigation that Zwicky had pioneered for fear that he might take offense and come after them. Still, astronomers did measure the so-called virial masses of clusters from time to time over the next few decades (the virial theorem is the one by which scientists can deduce masses from motions in systems that have reached gravitational equilibrium). Zwicky's bizarre result seemed to be consistent for one cluster after another. No one was sure what to do with this information; it was labeled the "missing mass problem" and filed in the section of astronomers' collective consciousness reserved for unsolved mysteries that might or might not be important.

Forty years or so after Zwicky's observation, an astronomer named Vera Rubin found herself involved in the same problem. Today a gray-haired grandmother, and still an active observer, Rubin, who is on the staff of the Carnegie Institution of Washington, is an anomaly among astronomers. She doesn't grab for all the telescope time she can get. Rubin reasons that two good observing runs a year are enough to provide her with good data but not so much that her life is seriously disrupted. And she prefers to avoid working on problems that are considered "hot." Even so,

Vera Rubin   PHOTO: EILEEN HOHMUTH-LEMONICK

she has more than once found herself making discoveries that have helped force the astronomical community into dealing with fundamental issues. Rubin is proof of the proposition that the most significant breakthroughs in astrophysics are often made not by those who are looking to make them but by skilled observers who have no particular theoretical axes to grind.

"When I first came to the Carnegie Institution in the 1960s," she told me during a visit to Carnegie's lush campus, "I was working on quasars, which had just been discovered. I didn't like it, though. It was too competitive. You'd get telephone calls from other observers who wanted to know your latest results even before you were ready to publish them, and trying to find out what you'd be looking at next. If you weren't going observing again soon enough, they'd want to rush in and do the observations themselves." Uncomfortable with this frenetic atmosphere, she abandoned the quasars and turned her attention to a topic nowhere near the frontiers of astronomy, but that interested her: the Andromeda galaxy.

Geoffrey and Margaret Burbidge, a husband-and-wife team based at the time at Kitt Peak National Observatory, were interested in Andromeda as well, and Rubin went to work with them for a year. One of the questions they wanted to answer was How fast does Andromeda rotate? It takes many millions of years for a large spiral galaxy to turn once on its axis, but astronomers can look carefully at the light coming from a spot on the galaxy's surface, and clock the speed of that spot. Rubin and the Burbidges did that, noting how fast different pieces of Andromeda were moving at the galaxy's leading and trailing edges, from the outside all the way into the core.

A galaxy looks like a single object, but of course it is not; it's a collection of billions of individual stars and gas clouds. Like the planets around the Sun, each of these objects rotates around the galaxy's massive core independently. And, like the planets, you'd expect that as you look farther and farther from the core, the stars and clouds would orbit more and more slowly. This kind of pattern is called Keplerian, since it was first discovered among the planets of the solar system by the German astronomer Johannes Kepler in the early 1600s (it's now known as Kepler's third law of planetary motion, but it applies to any object orbiting around a central mass).

The astronomers took all their measurements and drew a rotation curve, a chart that tracked the rotational speed as it changed from the center out to the edges. It wasn't Keplerian. "What we saw," said Rubin, "was that the rotation curve flattened out toward the edge." At a certain point, the rotation didn't get slower with distance from the core anymore; it stayed the same. "It looked strange," she told me, "but we didn't make a big deal about it, since this was only one galaxy, and the measurements were very difficult ones."

That was in 1970. In 1973, Jeremiah Ostriker and Jim Peebles, at Princeton, published a paper that concluded the Milky Way shouldn't exist. Its flattened spiral disk was inherently unstable, and the stars should long since have collected into a thick bar—unless, that is, the disk of visible stars were embedded in some sort of dark halo, roughly spherical in shape. The halo would resemble a glass paperweight with a butterfly—the visible Milky Way—at its center. Then Ostriker and Peebles, along with Amos Yahil, an Israeli astronomer, went on to write another paper that argued the halo was much bigger than the disk, and more massive as well. Besides stabilizing the disk, this halo would have the effect of flattening the disk's rotation curve out toward the edges—the same effect Rubin had already seen in Andromeda. "If you had been very smart," said Rubin, "you would have gone out right away and measured other spirals' rotation curves to test the theory."

By this test, she wasn't smart, but she found out Ostriker, Peebles, and Yahil were right anyway, in a more roundabout fashion. "By the middle of the 1970s," she said, "I had gone on to another subject that interested me." She wanted to know whether there were large-scale peculiar motions of galaxies, whether significant chunks of cosmos were moving with respect to the rest of the universe. "In fact, we did find that our galaxy, along with several others nearby did seem to be moving," she said. "It was very controversial, and nobody believed us. I found it all very unpleasant, and I went back to studying rotation curves again." Despite the initial and general disbelief, large-scale motions do seem to exist. They are the evidence the Seven Samurai used to infer the existence of the Great Attractor.

Working on rotation curves of several spiral galaxies with another Carnegie astronomer, Kent Ford, she said, "It didn't take more than a half-dozen exposures for us to realize that there was something going on. Every rotation curve we looked at was flat. Even then, people told us that this was an anomaly, that we were being fooled by looking only at the very brightest spirals, and that when we looked at dim ones the curves would start falling. They were wrong." She continued: "It's interesting—when you go back into the literature, it turns out that the first flat rotation curves were found in the late 1950s, with radio telescopes, but nobody paid any attention. I've asked people why that is. One reason seems to be that people simply *knew* that the rotation curves must continue to fall. In fact, you find statements to the effect that 'these observations show Keplerian curves.' But they don't! Another reason the observations went unnoticed is that they were in the radio, and there's a prejudice in favor of optical data. That's why things were settled relatively quickly after we published our data."

By the late 1970s, the missing mass problem had, by gradual consensus, become known as the dark matter problem. Missing mass was an inaccurate term anyway. The mass wasn't missing—it was there, but invisible. And it had begun to capture the attention of large numbers of astronomers. In particular, it caught the eye of George Blumenthal, a theorist at the University of California, Santa Cruz; along with his Santa Cruz colleagues Sandra Faber and Joel Primack and Cambridge astronomer Martin Rees, Blumenthal coauthored the papers that convinced many astronomers that cold dark matter was real. (Faber, whose office is just upstairs from Blumenthal's, was already known by then as a skilled observer with a firm grasp on theory as well. She went on to be one of the Seven Samurai, who discovered the Great Attractor.)

When I first went to see Blumenthal, in February of 1986, central California was going through the worst series of rainstorms in three decades. Mudslides kept blocking the major highways—the top layer of the landscape, including grass, bushes, and trees would slip down the hillsides intact, like skin being peeled back to expose the brown flesh underneath. The university itself is on a plateau, about five hundred feet above sea level, and two or three miles inland from the Pacific Ocean. As I drove up to campus though, a thick fog kept me from seeing anything farther than a half mile away. At the entrance, I thought I was heading onto a Scottish moor; all that was visible was acres of brown grass that covered rounded hills, set off by weathered wooden fences. The campus, less than thirty years old, was originally the Cowell Ranch, and the farm buildings are still preserved at the bottom edge of the range. But a mile farther on I drove without warning into a grove of enormously tall, straight trees—redwoods, I learned later—and if they were not technically giant redwoods, they were big enough to impress someone from the East Coast.

Blumenthal is tall, thinning hair on top balanced by a full beard; I never saw him wearing anything but a bulky sweater, well-worn corduroys, and running shoes, and he always looks just a little bit rumpled. He is also among the friendliest, warmest astronomers I've ever met. He is always welcoming when you show up at his door, always glad to talk on the phone (unless he's busy, in which case he schedules another time). He is willing to explain something twice, and then twice more. I once watched him at a press conference where he was on a panel talking about whether the Big Bang was in trouble, and bent over backward to address the hostile and persistent questioning of a reporter who clearly thought of mainstream astronomy as the province of charlatans and dupes. In fact he bends over, literally, when he stands or walks, slouching a little,

George Blumenthal

PHOTO: CATHY CLAUSEN

not (in my opinion) out of a sense of low self-esteem, but in an unconscious attempt not to overpower his companions.

The walls of Blumenthal's Santa Cruz office were plastered with turn-of-the-century bicycle-racing posters. "When I first came here," he told me, "my hobby was riding through the Santa Cruz mountains. Now my wife and I commute so much there isn't time, and I run for exercise." His wife, Kelly, teaches at Hastings College of Law.

Blumenthal told me a great deal about the rise of the cold dark matter model of the universe, virtually none of which I managed to absorb at first. Observers like John Huchra have a far simpler time than pure theorists at explaining what they do. They deal with big machines whose purpose is evident—it's easy to imagine photons of light banging into a broad telescope mirror and bouncing into a light detector; it's easy to visualize a galaxy, whose image is right there on a video monitor, flying away from Earth; and it's easy to look at the plotted positions of galaxies in a distance survey and recognize that they add up to a map. Theorists frequently deal with more abstract concepts, and the best they can manage in the way of a visual aid is often a chart that itself takes plenty of effort to understand.

Blumenthal is a pure theorist—so pure, in fact, that he has a horror of anything more mechanically complicated than his racing bike. Once, I happened to tell him a story about Ed Turner, whom I'd visited a few days earlier. While I was in Turner's office, a technician had arrived to deliver some extra memory chips for the computer. "Leave them on the table," Turner had said. "I'll install them myself." Blumenthal looked genuinely alarmed at the prospect of such a thing happening to him. "Not me," he said. "No way. I try to avoid such things. I don't even like software. I just want it to work."

I met with Blumenthal twice more; once in Cambridge, where he had gone on a year's sabbatical to work at the Center for Astrophysics, and again back at Santa Cruz, in February 1992. This time, the weather was spectacular, and I began to wonder how Blumenthal or anyone else at Santa Cruz had ever accomplished anything at all. From the open field I had first seen shrouded by fog, Monterey Bay took up perhaps a third of the field of view; it was lightly mist-covered, with mountains rising from the shore farther down the coast. In the foreground was a stand of eucalyptus, fully as tall as the redwoods, with immense, broccoli-like crowns. The redwoods were in the other direction, rising at the top end of the pastureland. It was a little warmer than average for February, Blumenthal told me when I got to his office—about seventy-five degrees.

Natural Sciences 2, the building where the physics and astronomy departments are, looked like a work in progress: vertical wooden beams, eight inches wide by sixteen inches deep, spaced about ten feet apart, were arrayed around the perimeter of the building on each floor, as though they were holding up the floors. They were. Santa Cruz was badly hit by the World Series earthquake of October 1989—the center of town, down near the ocean, was still closed off more than two years later—and Nat Sci 2 was the most damaged building on campus. "For three or four weeks after the earthquake," he said, "they didn't let us in at all. Then they let us in for ten minutes at a time, but only if we signed a waiver absolving the university of responsibility." The faculty was so upset that it insisted the university hire an outside engineering firm to certify the building's safety. The consultant ruled that Nat Sci 2, with the beams added, was just as safe as it was before the quake. "Of course," said Blumenthal, "look what happened then. But it didn't fall down. My attitude is, leave it alone. There's probably not going to be another earthquake this severe in my lifetime anyway."

Blumenthal's nonchalance was impressive, considering that he was in an elevator in the building when the quake hit. "It began shaking so hard I didn't have time to be scared," he said. "I was too busy trying to keep from being hurt. It was like being a bug shaken inside a jar. The really frightening part was during the aftershocks. They were mild enough so that I was able to think about the cable breaking. I remember trying to figure out the best position to be in—on my stomach, flat on my back, or standing up. I decided to lie on my back. Then, after about forty-five minutes, the lights went back on. I pushed 1, and the elevator calmly proceeded to 1." He used the stairs exclusively while we were together.

The CDM model of the universe took shape gradually during the late 1970s and early eighties, and many theorists were involved with its creation. Blumenthal was more involved than most, though, and so I turned to him to try and understand how CDM had come about—what observations and theoretical ideas and chains of reasoning had led astronomers into a theory that made so many odd claims about the cosmos. "Okay," he said. "I can't promise it will touch all the bases, but I'll give you my own revisionist version of history. I would argue," he went on, leaning back in his chair, "that modern cosmology began in 1965, with the discovery of the microwave background radiation. Starting around then, we began to get an idea of how the universe might have evolved—and to a first approximation, nothing has really changed since then." At the time, he was an undergraduate at the University of Wisconsin, Milwaukee. The next year he went on to the University of California, San Diego, as a grad student in astrophysics. The dark matter problem was already around, of course. "There was the Zwicky business, with the virial discrepancies. You saw galaxies moving around faster than you'd surmise from the gravity. I was well aware of the problem, and I even had a couple of crackpot theories, which I won't repeat. But people were thinking about it.

"The field didn't move a lot, though, until Peebles and Ostriker wrote their paper arguing that the disk of the Milky Way needed a spherical halo to stabilize it. But that wasn't convincing either. It was just another piece of evidence. One more piece was Rubin and Ford's rotation curves—in the midseventies they began going flat. I didn't work on the problem, though, because I really didn't have any good ideas. So I moved on to other things. I did my thesis on radiation processes, and I worked on X-ray sources, and, after I came to Santa Cruz, on quasars, too.

"But in 1979, Sandy Faber was asked to write a review article on dark matter. I talked to her before she began, and asked her whether she believed in it. She said no. A year later, after she had done the research, I asked her again, and she had turned around. 'You'd better believe in it,' she said. 'It's really important.' Even if I hadn't known Sandy and trusted her opinion, I would have been convinced by her article. All the evidence pointed to a ratio of about ten parts dark matter to one part visible matter in the universe. Now, it's dangerous to be convinced of something in science, because if you're convinced, you want to work on it, and if you're wrong, you waste your time. I went to work on the theory of dark matter.

"It's funny, but in 1984 there was a meeting in Princeton on dark matter, and I was struck that the focus of the meeting was not What is it made of? but Is it really there? I remember a real sense of disappointment: here were all these really smart people, and they were trying to decide on something I thought was already settled. That wasn't really fair of me. What was key was that they were recognizing an important problem."

If the dark matter outweighed the visible matter ten to one, it was clear to Blumenthal, as it was to just about everyone else in cosmology, that this invisible stuff must play a crucial role in any process involving gravity on a large scale. The overall structure of the universe, which presumably has come about through the action of gravity, would necessarily depend a lot more on dark matter than on the meager amounts of visible matter astronomers once thought was all there was. The formation of individual galaxies, too, must be more of a dark matter than a visible matter phenomenon.

Galaxy formation was something Blumenthal had thought about for years. "Back in grad school, I'd been very interested in the problem of how galaxies can exist. Cosmologists know that baryons—the protons and neutrons that ordinary matter is made of—interact with electromagnetic radiation." That meant that if the structure of the modern universe was indeed built by gravity, the seeds of that structure—the primordial lumps of matter that gravity molded over billions of years into the modern universe— should have created shadows in the radiation left over from that time. The cosmic microwave background should have patches of slightly higher temperatures than average to mark the locations of higher than average matter density.

"We were talking about clusters and superclusters of galaxies at that point," said Blumenthal, "since more exotic things like voids and bubbles hadn't been discovered yet. Based on the sizes of the structures we could see, we calculated backward to see how big the temperature fluctuations should look in the microwave background. It turned out that the fluctuations would have to amount to about one one-thousandth of the average temperature." They already knew, though, from observations, that the temperature differences were lower than that. "It was clear that in a simple baryonic universe, there hadn't been enough time to make the clusters and superclusters, given the small value of the microwave fluctuations. That puzzled us."

One other thing that puzzled Blumenthal, in particular, was that there were no objects in the universe that corresponded to any natural mass scale. Look out into space beyond the Milky Way, and while you see clusters and superclusters of galaxies, what is most obvious and striking is the tens of billions of galaxies, weighing in at somewhere between one hundred million and one trillion times the mass of the Sun. "So you ask yourself," said Blumenthal, "given the hot Big Bang, what is the characteristic mass of the first objects that should form in the universe? In a universe made of baryons, there are several such scales, and none of them corresponds to the mass range of galaxies."

These natural scales, he explained, are set by the laws of physics, and they reflect a balancing act between the tendency of gravity to make objects collapse and tendency of other forces to keep them from doing so. A similar sort of thing happens in the Sun: the heat and pressure of thermonuclear fusion at the Sun's core keep what amounts to a gigantic ball of hydrogen gas from collapsing under its own weight, and the battle between these two opposing forces sets the size of the Sun exactly where it is. In the early universe, the force that would have opposed gravity was the outward pressure of light itself, so hot and dense and energetic that it was able to blow apart any object smaller than ten quadrillion times the mass of the Sun. As the universe cooled and thinned out, perhaps a million years after the Big Bang, the natural scale would have dropped to one hundred trillion solar masses—still too big to be a galaxy. Finally, in the modern universe, the scale at which gravity and internal pressure are balanced is about a million times the mass of the Sun—in the range of the globular clusters of stars that orbit most galaxies, but too small for galaxies themselves.

"For both these reasons," said Blumenthal, "it seemed clear that there had to be a lot of nonbaryonic matter in the universe." The universe must, in other words, be made in significant part of something besides protons, neutrons, and electrons, some sort of matter that would not respond to the outward pressure of light and which could thus begin to collapse under gravity even when the cosmos was hot and dense. This nonbaryonic matter's indifference to light would also explain how it could, once it began to collapse into blobs of higher than average density, cast only the lightest shadow on the microwave background.

There were, as it happened, other reasons to believe in nonbaryonic dark matter as well, and at least two of them were tied in with a fundamental cosmological number known as omega. The idea is simple: the universe has been flying apart since the Big Bang. It is flying apart still. Will the gravity of everything in it—galaxies, gas, dust, dark matter—ever halt the expansion? Will the universe slow gradually to a stop, like a ball thrown straight up, and reverse course? It all depends on how much matter there is. If there's enough, omega is said to be greater than one, and the universe will eventually recollapse. The Big Bang at the beginning of time will be mirrored by a Big Crunch at the end. If there isn't enough matter to halt the expansion—if omega is less than one—then the universe will go on getting bigger forever, even if all the stars and galaxies in it burn out. And if—though it might seem so improbable that it's not worth considering—omega is precisely one, then the universe will always expand, but at a slower and slower rate. It will never quite stop, but its growth will be, after eons, almost imperceptible. (Technically, omega is the ratio between the density of matter there really is in the universe and the amount it would take to slow the expansion forever but never stop it.)

For at least a decade before George Blumenthal dove into dark matter in earnest, many astronomers were convinced that, improbable as it seems, omega is precisely one. In fact, it is *because* an omega of one is so improbable that it must be true. The reasoning was first trotted out in 1969 by Robert Dicke, a Princeton physicist. Omega, he pointed out, could have had any value at all. It could have been .000001 (which would have made the universe so thinly spread with matter that it would never have condensed into stars and galaxies). It could have been 1,000,000 (making the universe so dense that it would have lasted far less than a second before the Big Crunch. What omega appeared to be, in fact, was about .01, based on a rough estimate of the number of galaxies in the universe and the presumed weight of each. That was an unbelievable coincidence. Omega could take any value it wanted, and yet it was only a factor of one hundred away from being exactly one. How could this be?

One answer, proposed by Dicke, invokes what is known as the anthropic principle. Since values of omega far from one in either direction would mean the universe as we know it wouldn't exist, the fact that it does exist implies that omega must be close to one, improbable as that might have been. But a more popular line of reasoning said that omega is so close to one that it really is exactly one, and that we simply haven't found the missing 99 percent.

There was an even better argument: the laws of gravity dictate that if omega started out less than one right at the beginning of the universe, it would have kept dropping, and by now, billions of years from the beginning, it should be vanishingly small. If omega started out at more than one right at the beginning, it should have continued to grow, and now should be enormously large. The fact that it is .01 now implies that in the very first fractions of a second after the Big Bang it was .99999999999999999999999999999999999 99999999999999999999999. That seems more than a little improbable to most physicists.

This peculiarity of Nature is called the flatness problem since, in the language of general relativity, as created by Albert Einstein, gravity is more naturally explained as a bending of the shape of space and time than as a force between objects. A universe that collapses on itself—a universe with omega greater than one—is said to be closed; a universe that expands forever, with omega less than one, is open; and a universe with omega equal to one, poised just on the knife edge between closed and open is described as flat. The flatness problem asks: why is the universe so maddeningly close to being flat? The answer most popular with theorists is, again, that the universe is exactly flat, and that the extra mass that makes it so has not yet been found.

With the rotation curves of Rubin and Ford, the theoretical arguments of Peebles and Ostriker, and the measured speeds of galaxies within clusters, though, some of it had been found. The dark matter discovered by hard-nosed observers and analyzed by theorists had boosted omega from .01—1 percent of the way to flatness—up to .1, or 10 percent of the way. Now, if the flatness argument was to be taken seriously, astronomers were only 90 percent short of the most theoretically reasonable figure for the mass density of the universe. If the original figure of .01 was already improbably close to perfect flatness, the new number, ten times closer, was ten times more compelling. Furthermore, if observers had already managed to find a tenfold increase in the universe's mass density, it wasn't so hard to imagine that they could do so once again.

Like the original "missing mass," this new quota of matter needed to flatten the universe was obviously dark: it had never been seen. And even though its gravitational influence had never shown up in the rotation curves of galaxies or in the motions of galaxies through clusters, it could still be made of the same stuff that did show up. This extra component could be spread so evenly through the universe that its gravity pulled equally in all directions and thus didn't move things around. But what exactly was the dark matter? Was it just ordinary matter in a form too dim to be seen? Within the Solar System, there is a thousandfold gap in size between the Sun and Jupiter; maybe there are trillions of objects spread through the Milky Way and most other galaxies that fit somewhere inside the gap—objects too small and dim to be called stars, but too massive and hot to be called planets. Theorists who specialize in star formation even have a name for these objects: brown dwarfs. They wouldn't be visible more than a handful of light-years away, but if there were enough of them, they could add up to a respectable amount of matter.

It probably wouldn't be enough, though, to account for the dark matter that makes galaxies spin too fast and whip around at unacceptable speeds within clusters, and it certainly wouldn't be enough to bring omega up to the point where it would flatten the universe. That is not true only for brown dwarfs, either; it is true for any matter—asteroids, gas clouds, rocks, disposable diapers—made out of baryons. The same kind of stuff, proton- and neutron-based, that couldn't account for the smoothness of the microwave background and for the mass range of galaxies, also couldn't account for enough dark matter to flatten the universe.

To understand why, I nearly had to go to Chicago, which is not especially nice in February. Fortunately and unexpectedly, I found out that Chicago had come to the San Francisco Bay Area, in the form of David Schramm. Schramm is a cosmologist, too, and a theorist, but of a different sort than Blumenthal. Instead of considering the galaxies and stars, Schramm and the group of scientists he has assembled at the University of Chicago and the Fermi National Accelerator Laboratory nearby, concentrate on subatomic physics. The idea is that, in the earliest minutes after the Big Bang, the first things that formed out of the pure energy of the initial blast were not planets or stars or even grains of dust, but elementary particles. The particles then led to everything else and are still around, in one form or another. Study them, and you can learn a lot about the early universe.

David Schramm had come west to spend six months at the University of California, Berkeley, where astronomers and physicists had recently set up the Center for Particle Astrophysics to investigate the cosmic-subatomic connection. As a sort of elder statesman of this branch of cosmology (he began churning out important papers nearly twenty years ago), he was there to help get the research started.

I had not seen Berkeley since I was seven years old, and at the time it had not occurred to me that the school resembles the early universe in many ways. It was hot and terribly dense, with what appeared to be several million students packed into the plazas and

walkways of the hillside campus. There were pockets of turbulence, as knots of even higher density formed around card tables
pushing every imaginable political, social, or ethical view. The
biggest eddy swirled around an impeccably clean-cut young man,
carrying a Bible under one arm, who had clearly been interrupted
in a sidewalk sermon and was arguing with another young man
who was not so clean-cut. "Oh, yeah," said the heckler, "well, if
God can do anything, can he create an object so heavy that He
can't lift it?" As the preacher tried to answer, he was shouted
down by another heckler. "Hey, free speech, man," shouted someone else in the crowd.

I found Schramm at a spare desk that had been crammed into
a corner of one of the center's offices. He looked crammed in as
well; he is both tall and wide, a former Greco-Roman wrestler,
with curly blond hair. He took a break from his equations, clasped
his hands behind his head, and leaned back in his chair to explain

David Schramm   PHOTO: COURTESY UNIVERSITY OF CHICAGO

why the dark matter can't be made of baryons. It can't because lithium and other light elements—elements like hydrogen, deuterium, and helium—say so. "The relative proportions of the light elements," he said, "tell you that at one time, the early universe was a hydrogen bomb, converting hydrogen into helium at very high temperatures. What's so impressive is that we can tell you not just what the abundances of the major elements, like hydrogen and helium, should be, but also of very scarce elements like lithium. The microwave background tells you about the universe when it was a few hundred thousand years old. The light elements tell you what the universe was like when it was one second old."

What Schramm and his Chicago collaborators did in the early 1970s was to take the hot Big Bang as a given, and look in detail at the nuclear physics expected to occur in the first few seconds when temperatures were hovering at around ten billion degrees and pressures were millions of times greater than they are anywhere, even in the cores of massive stars, today. The physics told the scientists that, before it cooled off and thinned out, the universe should have cooked itself into a proportion of about 75 percent hydrogen, almost 25 percent helium, .01 percent deuterium (a form of hydrogen whose nucleus consists not of a single proton but of a proton and a neutron), and .00000001 percent lithium. "We predicted these abundances," said Schramm, "and then the observers went out and looked at the oldest stars, which contain primordial matter, and found that the abundances are just what we predicted."

The Chicago group then went on to make another prediction. Physicists had decided by the end of the 1960s that protons and neutrons (and a number of other particles as well) were themselves made up of smaller units: the quarks. These appeared to come in at least three separate "families" (which are characterized in ways too complicated to discuss here). It was an open question whether there were more families than that—a loose end that made everyone unhappy. Schramm and his crew found a way to tie it up. They used another elementary particle, called the neutrino, a particle that has virtually no mass at all (or possibly even has no mass, period), no electric charge, and zips through matter as though it weren't there. Build a wall out of lead, dozens of light-years thick, and the typical neutrino will breeze right through. It turns out that there are different types of neutrinos, too, and that the number of types is intimately connected with how many families of quarks there are. Three families of quarks means three types of neutrinos, four means four, and so on.

It also turns out that the amount of helium created in the early universe depends on how many types of neutrinos there are. So Schramm and the Chicagoans took the fact that helium adds up to a quarter of the baryonic mass of the universe, worked the equations of nuclear physics, and came up with the conclusion that there are at most four types of neutrinos, no more, and probably only three. That prediction came out in the mid-1970s. In 1990, the Large Electron-Positron accelerator, on the border between France and Switzerland, owned by the *Conseil Européen pour la Recherche Nucleaire,* or CERN, found evidence that there are, in fact, three types of neutrinos, and almost certainly no more. "That was the first time in history," said Schramm, "that anyone used astronomy to predict what physicists would find in a particle accelerator."

It was the relative proportions of light elements, though, that proved baryons could not account for dark matter. "What the light element abundances tell us," said Schramm, "is that if you're talking about baryons alone, there is at most ten percent of what you would need to close the universe." That might be enough to account for the motions of galaxies; the amount of dark matter inferred from its gravitational influence on what we can see is about 10 percent of what it takes to close the universe as well. But maybe it isn't quite enough. Some observers think there is direct evidence for two or three times that much dark matter—enough so that something other than baryons would have to make up at least some of it. And if the flatness argument is correct, the need for nonbaryonic dark matter is even greater.

The size of galaxies hinted to astronomers that there must be a lot of nonbaryonic matter in the universe. The lack of a significant imprint on the microwave background radiation implied the same thing; and the amount of dark matter found by people like Rubin and Zwicky, along with the maximum amount of baryonic matter permitted by the universe's helium content, did, too. "If you thought about it at all," said Blumenthal, when I returned to Santa Cruz, "it made a lot of sense that the dark matter had to be made of something other than baryons. Nonbaryonic particles could start clumping under gravity before decoupling, and the photons blasting through the universe at that time wouldn't tear the clumps apart. After decoupling, the mass concentrations would already be there, and baryons could fall into them. And if ninety percent of the universe is dark matter, the fluctuations in the microwave background are suddenly ten times less pronounced." In an all-baryon universe, the fluctuations should have been seen by then; in a mostly nonbaryonic universe, they shouldn't. In the real universe they weren't.

"The first papers I remember seeing suggested that neutrinos were good candidates for the dark matter," said Blumenthal, "although in general any weakly interacting particle would work." Neutrinos did have a few advantages over other possible candidates. For one thing, they were known to exist, having been discovered in the 1950s. For another, there are plenty of them. Trillions pass through the average human body, utterly unnoticed, during the average second, and they don't interact much with matter or radiation. One problem, however, with proposing that neutrinos were the source of nonbaryonic dark matter was that physicists had been assuming for years that, unlike the protons and electrons that make up atoms, but just like the photons that carry electromagnetic energy, neutrinos had no mass at all. No mass means no gravity, and no chance that they could be responsible for making galaxies spin too fast or for halting the expansion of the universe.

In fact, though, no one had ever *proven* that neutrinos have zero mass—only that their mass is incredibly small. It would have to be. Given their numbers in the universe, neutrinos that weighed as much as, say, an electron, the lightest particle known, would have added so much mass to the universe that the Big Crunch would have happened long ago. If they were going to make omega equal to just exactly one, neutrinos would have to weigh no more than about .0006 percent as much as electrons, no more than thirty electron volts in the particle physicists' system of weights and measures.

In 1980, two different experiments, one in Moscow and one at a reactor at the U.S. government's Savannah River nuclear weapons installation, seemed to point to just that value for the neutrino's mass. The latter, run by Fred Reines, who had discovered the neutrino a quarter century earlier, hinted that neutrinos in at least one of the three known families could change identities, spontaneously joining another family. If he was right, all of the neutrinos had mass, although it couldn't be pinned down. The former experiment did pin down the mass: the Moscow physicists claimed that the mass of the neutrino, deduced from the radioactive decay of tritium, was less than forty-five and more than eighteen electron volts.

Both experiments were extraordinarily delicate and the chance of a mistake was reasonably high, but the results added to a growing sense that neutrinos were going to solve the dark matter problem and the microwave background problem all at once. It didn't make Blumenthal happy. "The problem with neutrinos," he said, "is that they would tend to create enormous structures." When baryons try and form into a blob, there are two competing forces at work: gravity, which makes the blob want to collapse, and pressure, which resists collapse. For weakly interacting particles like neutrinos, there is no such thing as pressure, since the particles don't even notice each other, let alone other matter or light. But they do have inertia; the particles that condensed out of the energy of the Big Bang were born moving and they would tend to stay in motion unless forced to do otherwise. "The characteristic scale of structure for these particles," explained Blumenthal, "is determined by the time it would take a blob of particles to collapse under gravity versus the time it takes the typical particle to traverse the blob." Unless a blob is big enough, a particle would escape before gravity has a chance to stop it.

The tendency to keep moving is known as free streaming, and the size of the smallest blob that can form is called the free-streaming scale. For neutrinos, which move at or near the speed of light, the free-streaming scale is between one and ten quadrillion times the mass of the Sun. Anything smaller would be washed out; by the time the blob collapsed, there would be no neutrinos left inside. If neutrinos were indeed the dark matter, then it meant that the most fundamental mass concentrations in the universe should be four orders of magnitude—ten thousand times—bigger than galaxies. Presumably, the blobs of neutrinos would fragment as they collapsed under gravity, breaking into miniblobs of the size of superclusters, which would break down to form clusters, which would finally fragment into galaxies. But the gigantic blobs would form first.

"It meant," said Blumenthal, "that we're a long way from understanding the problem of galaxy formation, which was one of my central interests. I found it depressing. That didn't mean it wasn't true; the universe is not obliged to be simple, but I didn't like it." Like it or not, there were hints that it might be true. Although the Great Wall hadn't been discovered yet, observers had begun to see indications that matter—galaxies and clusters of galaxies—were organized on a larger scale than anyone had suspected. There was even one theoretical camp, whose leader was Yakov Zel'dovich, in Russia, that argued that the universe had evolved from the top down, from enormous clouds of matter that fragmented into smaller and smaller sizes. The idea of neutrino dark matter made that group as happy as it made Blumenthal uncomfortable.

Fortunately for Blumenthal, it didn't take long for the neutrino idea to go out of fashion. One reason was a theoretical paper by

Scott Tremaine and Jim Gunn, both at Caltech during the early 1980s, now at the Canadian Institute for Theoretical Astrophysics, in Toronto, and Princeton respectively. What the paper pointed out was that elementary particle physics forbids neutrinos to be packed together too tightly. The less massive they are, the harder it is to pack them together. Considering how massive the dark halo surrounding the Milky Way is, and considering how tightly neutrinos can possibly be packed, the least they could weigh is thirty electron volts. That's also the most they could weigh, if omega really is one. "So it was just barely possible," said Blumenthal, "but so close that it made people very uncomfortable." Furthermore, because it wasn't possible for the neutrinos to be packed tighter, it meant that galaxies much smaller than the Milky Way shouldn't have significant dark halos at all. "But observations of the dwarf galaxies in the Local Group show that they do have massive dark halos," said Blumenthal. "So either dwarf galaxies have a different kind of dark matter than big spirals, which is very hard to believe, or the halos are not made of neutrinos."

The other evidence against neutrinos came from the only sort of experiment astronomers can perform on a universe that is otherwise impossible to manipulate: a computer simulation. There is only one real universe, but a scaled-down cosmos whose history unfolds inside a computer can test out all sorts of theories and compare the results with the real thing. If the two don't match, then the theory is a bad approximation of what is going on in the real universe. With neutrinos in the computer, the imaginary and the real didn't match. A neutrino-dominated universe does indeed form structure on the largest scales first, and on the scale of galaxies last. This means that the galaxies should have formed relatively recently. In fact there is strong evidence that galaxies actually started forming when the universe was less than half its present age.

Astronomers began casting around for another particle they could use. There weren't any known to exist that worked very well. But at about the same time that George Blumenthal was thinking about how to make galaxies and trying to understand why there was no detectable temperature variation in the cosmic microwave background, his colleagues in elementary particle physics were playing with a group of new intellectual toys known as grand unified theories, or GUTs. As far as anyone knew, there were four ways matter could interact with other matter. There was electromagnetism, which binds electrons and nuclei together to form atoms, atoms to form molecules, and molecules to form matter. Not surprisingly, electromagnetism is responsible for electricity and magnetism as well, and it's the force behind light in all its forms. There was the strong nuclear force, which makes quarks into protons and neutrons, and these particles in turn into atomic nuclei. There was the weak nuclear force, which is the cause of certain forms of radioactive decay (and which is the only force neutrinos experience, besides gravity). And there was gravity.

But the idea of four separate forces seems uneconomical, and physicists have long been convinced that these evidently different phenomena are, at the core, the same. It's only in the modern universe that they seem different. In the stupendous heat of the Big Bang, they were unified, indistinguishable. The search for a single, unified theory of everything, which would demonstrate the equivalence of the forces at high temperatures and explain their differences at low, is a major program of modern physics, and by the late 1970s the theorists had taken the first step: they showed that electromagnetism and the weak force were once unified, and hadn't separated until sometime during the first second of the universe's life. (Strictly speaking, this was really the second step, since James Clerk Maxwell, the Scottish physicist, had proven back in the midnineteenth century that electricity and magnetism are different manifestations of a single underlying force. But this happened so long ago that it is considered ancient history, and is usually left out of discussions about unification.)

Even as dark matter was becoming a hot topic, physicists were working on the next step: finding the underlying identity of the newly understood "electroweak" force with the strong nuclear force. The fledgling theories they were working on were known, collectively, as GUTs. (Physics is evidently not immune from title inflation. If the tying of three out of four forces together is considered "grand," what will it take to combine all four? The term has already been invented: superunification.) There were, and still are, plenty of versions of grand unified theories around, since no one could figure out which, if any, was correct. One common feature seemed to be that most of them predicted the existence of new, undetected elementary particles. In one version, called supersymmetry, every known particle would have a corresponding supersymmetric particle. If the theory is right, explained Blumenthal, "then a lot of these particles will interact as weakly as neutrinos, and many will have decayed since the Big Bang into lighter particles." The question was, what is the lightest supersymmetric particle, and in the early eighties, the answer seemed to be the gravitino, with a mass of about one kilo-electron-volt, or one five hundredth that of the electron.

"We asked ourselves, 'What is the free-streaming mass of the gravitino?' and calculated that the answer was ten to the eleventh solar masses. We said, 'Hey, that's the mass of a galaxy.'" They had finally found something that could build galaxy-size objects naturally. "That was really exciting," said Blumenthal, "but our excitement didn't last long. It turned out the gravitino wasn't the lightest supersymmetric particle after all. And when we thought about it, we realized that while gravitinos naturally clumped on the scale of a large spiral galaxy, that couldn't explain why there are so many small galaxies, with only one hundred million or so stars."

The neutrino is an example of what astrophysicists call a hot particle; it is very light, if it has any mass at all, and as a consequence, when it's born (whether out of the Big Bang or out of the decay of a neutron) it is born moving at or close to the speed of light. That is what gives it such a high free-streaming mass; a blob has to be big if it's going to be too big for a light-speed particle to escape from it. This tendency to boil off into space is the reason it's called hot; Dick Bond, a theorist now working with Tremaine at the Canadian Institute for Theoretical Astrophysics, gave it that label. Gravitinos are more massive, and slower; they're considered to be "warm" dark matter.

"It was kind of obvious," said Blumenthal, "that if you're thinking about hot dark matter and warm dark matter, and they don't work out, you should investigate cold dark matter." The first person who described cold dark matter (CDM) was Jim Peebles. "He wrote a paper just presuming that this stuff existed," said Blumenthal, "and calculated the effect on the microwave background. Then Dick Bond and Alex Szalay looked at a similar idea, and the Gang of Four—Simon White, Carlos Frenk, George Efstathiou, and Marc Davis—started to do large-scale computer simulations to see how cold dark matter would influence the evolution of the universe. My own motivation for getting involved in CDM was because it was the obvious way to go."

Meanwhile, the grand unified theorists had also come up with some new particles to play around with. They were cold—though, like the gravitino, neither had been seen anywhere outside of an equation. One was the axion, which was named, perversely, after a laundry detergent. The other was the photino, the supersymmetric counterpart of the photon. The photino, at a billion electron volts, was (and remains) the replacement for the gravitino as the lightest supersymmetric particle. The axion didn't have a well-understood mass, but it wasn't impossible to imagine that it could have a mass comparable to the photino's. "That was another motivation to get into CDM," said Blumenthal, "the fact that these candidate particles were being discussed."

By then Blumenthal was working closely with Sandra Faber and Joel Primack. There were two real questions, they agreed. First, what were the initial fluctuations that led to the eventual growth of structure in the universe? And second, how did the universe modify those fluctuations up until the time of recombination and decoupling, when, presumably, gravity took over and began acting on its own?

The initial fluctuations had always been a major assumption of astronomers who were trying to understand structure in the universe. You can't make something out of nothing. You can't make galaxies, superclusters, and blobs of neutrinos weighing as much as ten quadrillion Suns if there isn't some sort of structure in the very first moments of the universe to work with, some molehills out of which to create mountains. Given just the merest hint of density variation from one spot to another in the early Big Bang, gravity would have automatically amplified that variation, pulling matter from the sparser regions into the denser, thus making them denser still. (Electromagnetic radiation, pressure or fast particle movement will, of course, erase any fluctuations that aren't massive enough, but the biggest ones will survive.) If there hadn't been any density fluctuations at all, the universe today would have no galaxies, and no astrophysicists to complain about it. It's another version of Dicke's anthropic principle.

"The idea of initial fluctuations had been around for a long time," said Blumenthal, "but no one had very good ideas about them. People thought, for example, that they might come from turbulence in the Big Bang. But turbulence disappears quickly if you don't keep pumping in energy, so that didn't work very well: there was nowhere for the energy to come from." Then, in the early seventies, there were papers by Zel'dovich, Ed Harrison, of the University of Massachusetts, and Peebles that attempted to set some sort of limits on the fluctuations. Because they would have been generated by a random process, the fluctuations would almost certainly have come in a variety of both sizes and intensities.

"The reasoning went like this," said Blumenthal. "On very small scales, the fluctuations can't have been enormously dramatic." If they had been, the areas of very high density would have affected the process of nucleosynthesis, the creation of the light elements whose characteristics Schramm and the Chicagoans had calculated so carefully. Since their predictions matched the real universe so well, said Blumenthal, "that sets a limit on how much fluctuation there could have been at these scales. On the very largest scales, there's an upper limit to the intensity of fluctuations as well: too big, and they would have left an imprint on the microwave background. And finally, the structure we see today (again, we were talking only about clusters and superclusters at that point) sets a lower limit to the fluctuations. They must have been big enough, at least, to create what we now see."

Hemmed in by these three limits at different scales, Harrison, Peebles, and Zel'dovich showed mathematically that whatever the cause of the fluctuations might be, they can't have had any characteristic scale. The degree of fluctuation must have been just about the same as large, medium, extralarge, and tiny scales. "This isn't physics, though," said Blumenthal. "It's phenomenology—fitting the theory to the available data. What we really wanted was a physical basis for the Harrison-Zel'dovich-Peebles spectrum, a reason to *expect* that spectrum and no other, rather than the simple observational evidence that that's what happened.

"Then inflation came along and gave us a reason."

It did more than that. The theory of inflation, which burst into the collective consciousness of astrophysicists in 1980, not only explained the Harrison-Zel'dovich-Peebles spectrum but also solved the flatness problem, explaining why omega would have to be equal to one. It solved the problem of why there are no monopoles in the universe, particles with just a north or a south magnetic pole but not both, when grand unified theories suggested they should be all over the place. It even solved another longstanding mystery of cosmology known as the horizon problem. The horizon problem had to do with the uniformity in all directions of the microwave background radiation. Look in one direction, say, straight up from the north pole. The background radiation is humming in at 2.7 degrees above absolute zero. Now look straight up from the south pole. Same temperature, precisely. The problem is that the microwave radiation from the north has been traveling since it was emitted soon after the Big Bang; it has been traveling for perhaps fifteen billion years at the speed of light. Same with the radiation from the south. Here is the horizon problem: if nothing can travel faster than light, and if these two opposite edges of the electromagnetic universe are each fifteen billion light-years from us, then they're thirty billion light-years from each other. There's no way that microwaves or anything else—including heat—can have gone from one side to the other in the age of the universe. And if the two sides of the universe can't have transferred heat to each other any time during the history of the universe, there's no way other than bizarre coincidence that could have let them be at precisely the same temperature, and physicists are very suspicious of bizarre coincidences. (It would be tempting to argue that in the first moments of the Big Bang the two sides of our currently observed universe were a lot closer together and could have exchanged heat with each other, but it doesn't work because radiation from each side, traveling as fast as it possibly could, has just now gotten to us.)

Inflation solved all these problems in physics and astrophysics, and all its proponents asked scientists to do was accept the fact that between about .0000000000000000000000000000000001 second and .0000000000000000000000000000001 second after the Big Bang, the universe expanded by a factor of 1,000,000,000,000, 000,000,000,000,000,000,000,000,000, going from the size of a proton to the size of a grapefruit or thereabouts. After that, it went back to its regular rate of expansion. That's a very short time—but such a dramatic growth rate that the universe was expanding—for that brief interval, at much more than the speed of light. This doesn't violate any of the rules of relativity, which put a speed limit on things traveling through space but not on space itself. And it makes all of the troubling cosmological problems go away at once.

Take the horizon problem. The opposite sides of the visible universe were in close contact early on after all. It's just that they were whisked apart at warp factor one thousand after they had shared heat and equalized temperatures—whisked so far away that only now has their microwave light finally made it back into our neighborhood. Take the flatness problem. Recall that in the language of general relativity, flatness of space and a density of matter that will just fail to recollapse the universe are interchangeable concepts. Imagine that you're balancing on top of a (very strong) beach ball. It's obviously curved. Now let it inflate to the size of the Earth. It's still curved, but as far as your eyes can tell, it's perfectly flat. Same idea with the universe: it may indeed be curved at some scale far beyond the horizon, but the part we're floating in has been inflated so much that all we can see is flatness. Take the monopole problem. There were lots of them around in the early universe, but things have blown up so much that they're now spread thin. In all of our visible universe, there should be approximately one, and therefore it's no surprise that nobody has yet found it.

Inflation is, in short, a beautiful theory, a single, straightforward solution to several evidently unrelated and nagging problems in astrophysics. And while it might have been dismissed as purely nutty—after all, if you can just speed up the expansion of the universe to make things fit, why not make it jump through any number of other hoops—inflation had a reasonably solid basis in physics. The phenomenon emerges naturally from the same grand unified theories that give plausible existence to monopoles and axions and gravitinos and photinos. It was discovered by a postdoctoral student who had no formal training in grand unified theories or in astrophysics—he heard of the flatness problem in a lecture at Cornell by the visiting Bob Dicke and was impressed with the idea but had no thought of pursuing it. Alan Guth was a particle physicist who, after graduate school at Princeton and postdoctoral jobs at Columbia and Cornell, had signed up for yet another postdoctorate at the Stanford Linear Accelerator Center (a few dozen miles north of Santa Cruz). It was there, working at home late one night, that he discovered the secret of the universe.

One fundamental consequence of grand unified theories is that they imply all sorts of exotic particles. Another is that as the universe cooled, the emergence of individual forces—the strong force, the weak force, electromagnetism—was accompanied by a fundamental change in the underlying energy structure of the universe, a phase change analogous to what happens when water freezes into ice. The thing that actually changes is a universe-wide energy field called the Higgs field (the Higgs particle, the theoretical carrier of Higgs energy, is something physicists had hoped to discover with the eight-billion-dollar superconducting super collider, the particle accelerator fifty-three miles in circumference that was being built around the town of Waxahachie, Texas before Congress killed it in 1993). The breakdown of grand unification and the freezing-out of the Higgs field happens at about 10^{27}, or a billion billion billion degrees if everything goes right.

What Guth realized was that everything might not go quite right. When water freezes, ice first appears around impurities, as pearls form around grains of sand. Then the rest of the pond or the ice cube tray follows. But if the water is exceptionally pure, and treated gingerly, it can actually be drawn well below the freezing point and remain liquid. The process is called supercooling. Eventually, some microscopic eddy of turbulence will act as an impurity, and the water will finally freeze, releasing lots of pent-up heat all at once.

The same thing can happen with the Higgs field. At ten to the minus thirty-five seconds and 10^{27} degrees, Guth found, the universe could have become supercooled, dropping below the point where the forces should have frozen out without their freezing. The universe entered a state of what Guth labeled "false vacuum." It was full of nothing, but the nothing was bursting with energy. This energy manifested itself as antigravity, with a force that grew stronger as the nugget of false vacuum grew bigger, which it did at a confounding rate. Finally, when it had grown from a proton into a grapefruit, the Higgs field froze at last, the expansion slowed down to the original relatively leisurely pace of the Big Bang, and the energy stored in the Higgs field was released in a burst of particles. Guth figured that to account for all of our currently visible universe, about twenty pounds' worth of false vacuum would do nicely.

In addition to solving the horizon problem, the flatness problem and the monopole problem, inflation gave Blumenthal and the other dark matter theorists a natural way to generate the fluctuations that would eventually turn into the structure of the universe. Like any energy field, the Higgs field presumably has to obey the laws of quantum physics, which govern the behavior of particles and energy at microscopic scales. One of those laws is the Heisenberg uncertainty principle. It asserts (and experiments have repeatedly proven) that on small enough scales and over short enough time spans, it is impossible to characterize the precise position of a particle or the energy of a particular region of space. It's meaningless even to talk about there being such an exact position or energy. The result is that the Higgs field can't have been smooth at the submicroscopic level, but seething with random fluctuations—fluctuations, as it happens, with a spectrum just like that of Harrison, Peebles, and Zel'dovich. When the universe inflated, that spectrum of fluctuations would have been blown up, too, made macroscopic, and thus been available to impose ripples on whatever particles or radiation filled space.

"So when we were all playing this game in the early eighties," continued Blumenthal, "we could look to two different reasons to believe in the spectrum: the phenomenological reason and the reason that came out of physics." Armed with the spectrum of initial density fluctuations—which were simply presumed before the discovery of inflation, and deduced as a consequence of inflation afterward—Blumenthal, Faber, and Primack then concentrated on the second question: how did the forces in the early universe modify that spectrum? "In the hot dark matter model, for example," said Blumenthal, "the modification comes from free streaming by neutrinos, so that by recombination all the fluctuations smaller than ten to the sixteenth solar masses are gone. So we asked ourselves how cold dark matter would operate, and what should the implications be for galaxy formation?"

Because they aren't destroyed by free streaming or blown apart by the pressure of radiation, blobs of dark matter particles, mirroring the random fluctuations in the underlying energy fields of the universe, were free to begin collapsing under gravity a few minutes after the Big Bang. "We made the assumption," said Blumenthal, "that the blobs that collapse are spherical. This is obviously not completely true, but we decided it was a reasonable simplification. This tendency to collapse under gravity was balanced by the expansion of the universe. The blobs were located in space, and space was expanding, so the blobs did, too—just a little slower than everything else. At a certain point, though, gravity wins and the blob really does collapse."

Cold dark matter particles are easier to collapse than neutrinos, and they can be packed closer together, but there is still a limit to how tightly. The limit comes from their inability to emit radiation and bleed off some of their energy of motion. Blumenthal began rummaging through a pile of paper to find an equation that could help make this point. "It's not that I can't reproduce the equations from my head," he said, "but actually, I can't. Anyway, it turns out that the collapse stops when the blobs have shrunk by a factor of two."

The calculations showed that particles of baryonic matter, now relegated to the status of an impurity in the dark matter universe, were still pretty well mixed in with the nonbaryonic particles at this point. "But unlike the dark matter," said Blumenthal, "the baryons can keep on collapsing." To do that, they have to radiate away their energy, so the question Blumenthal, Primack, and Faber asked was: how fast can they do it? The answer was that for blobs of between a million and a trillion solar masses, the baryons could shed their excess energy in the first tenth of the universe's lifetime. "That's just the mass range of observed galaxies," said Blumenthal, "and I remember looking at this result with a sense of awe. It gave a physical argument for the size of galaxies. That really made me think this was a serious idea. Today, I'd say it's a good idea for spirals, and maybe not so good for ellipticals, but spirals do make up most of the galaxies in the universe.

"Then we looked at this in a different way. For each galaxy

mass, we looked at what the velocities should be inside—the speed of rotation, for spirals, and the random motions of stars within the galaxies, for ellipticals. What we found was that the mass went as velocity to the fourth power. That, it turns out, is just about the same as the Tully-Fisher and Faber-Jackson relations." These two relations—the second had been discovered by Sandra Faber working with another astronomer named Robert Jackson—actually tie the velocities within galaxies to the galaxies' inherent brightness, not their mass. But since brightness comes from stars, and more massive galaxies have more stars, the relationship discovered in Blumenthal, Faber, and Primack's calculations seemed to amount to the same thing. Faber-Jackson and Tully-Fisher are among the most effective ways to estimate distant galaxies' inherent brightness, which, when compared with their observed brightness, gives astronomers an idea of how far away they are.

"So the theory had given us a physical basis for another phenomenon that was simply observed before. And there was one more example: the galaxies we formed using cold dark matter could be expected to have flat rotation curves."

There was only one obvious problem with the cold dark matter model, and it emerged when the model's initial conditions and equations were plugged into a computer and allowed to evolve in an imaginary, electronic universe. The problem was that galaxies didn't seem to cluster in the model as much as they cluster in the real universe. Huchra and Geller's distance survey was not yet complete at that point, and while there had been hints of enormous cosmic structures early on, no one knew how striking and widespread the structures were. But as early as the early 1970s, starting with Princeton's Jim Peebles, astronomers already knew that galaxies all over the sky were clustered together more strongly than mere chance would suggest. Peebles had come up with a statistical device called the two-point correlation function. What it said was this: given a single galaxy's position on the sky (its two-dimensional position, leaving aside distance), what is the likelihood of finding another galaxy within any given distance? Peebles had found that it was greater, for a wide range of distances, than pure random chance would have predicted.

In the early 1980s, another astronomer, Neta Bahcall, an Israeli who emigrated to the United States in 1966, took the statistics one step further. Bahcall is now at Princeton, but from 1982 to 1990 she worked at the Space Telescope Sciences Institute, in Baltimore. "I was lucky," she told me, sitting in a small office at Peyton Hall, the astrophysical sciences building that is overshadowed by Jadwin Hall, the university's huge physics building, just downhill. "The troubles with Space Telescope began just six months after I left."

It was in Baltimore that she and a colleague, Ray Soneira decided to test the two-point correlation function on rich clusters of galaxies, not just galaxies themselves. Rich clusters, so labeled by the astronomer George Abell in the 1950s, are those with fifty or more galaxies squeezed into a volume about three million light-

Neta Bahcall   PHOTO: EILEEN HOHMUTH-LEMONICK

years across—comparable to the distance between the Milky Way and Andromeda. "What we did is like mapping a mountain range by looking only at the very tallest peaks. You miss out on the detail, but you get a good idea of where the structure is. The clusters of clusters—the superclusters—are really the pillars of large-scale structure. And their existence means that individual clusters are much more highly correlated than individual galaxies are." Look at a galaxy, and you're likely to find another galaxy nearby. Look at a rich cluster, and you're very likely to find another one nearby.

It was this clustering, of galaxies and of clusters themselves, that didn't mesh with the cold dark matter model. The computer simulations made it clear that if galaxies had to form wherever there was matter, the cold dark matter universe would have less structure than the real universe astronomers could see through their telescopes. But maybe galaxies didn't have to form everywhere there was matter. Maybe they had a preference for forming only in the very densest regions of the universe. That was what the dark matter theorists proposed. They called the phenomenon biasing— the galaxies showed a bias in favor of high-density areas. The 90 percent of the universe that would bring omega up to one was hidden because it was a shade too low in density to have made galaxies. But there was a lot of it because it filled in the vast, seemingly empty spaces between the clusters and the superclusters.

"Biasing," said Blumenthal, "is an idea that Nick Kaiser [who is now at the Canadian Institute for Theoretical Astrophysics, along with Dick Bond and Scott Tremaine] first published, and many of us had been thinking of. Basically, it means that the distribution of light doesn't reflect the underlying distribution of mass. Initially, biasing was regarded as a very negative thing. In fact, it's probably inevitable, since it's obvious that some regions will be better at making galaxies than others. I have no idea of what the physical significance of biasing will turn out to be."

The theoretical significance, however, was obvious. When the Gang of Four and other keepers of computer-simulated universes put in the additional factor of bias, their universes finally began to resemble the real one. The fundamental pillars of the cold dark matter model were in place: omega equals one; inflation; dark matter is nonbaryonic, made up of weakly interacting elementary particles; and galaxies form with a bias toward high-density regions of the universe.

"In the early days," said Blumenthal, "a lot of things were working. I think many of them still do." What doesn't work anymore is the argument that biasing solves all the theory's problems. Biasing could explain some of the excessive structure seen in the first half of the 1980s, but it couldn't handle what astronomers found in the second half: the Great Wall, the Great Attractor, and even higher estimates of the galaxy-galaxy and cluster-cluster correlations. All point to a universe where the biggest structures are too big.

"The problem with CDM today is that there's still too much power on large scales," said Blumenthal. "The options are really threefold. First, maybe the universe has not just cold dark matter, but hot dark matter as well. Or maybe there are more baryons than we think. Either of these would help, but it would take a nasty God to give us two forms of dark matter. It would work, but it's ugly. Or maybe you could change the spectrum of fluctuations. They're all possible, but I don't much like any of them. You have to remember that when people say CDM, they usually mean standard CDM: omega is one, inflation, biasing of one point five to two point five, and a mass to light ratio of about ten to one. You can relax any of these and still have dark matter. If I wanted to believe in CDM, and you told me that omega is much less than one, it wouldn't necessarily destroy me, though it would make me less happy."

Not everyone in the astronomical community, though, was as convinced by the CDM theory as Blumenthal, even in the early days when the closer theorists looked the more questions the theory answered. Ed Turner, the Princeton astronomer who likes to install his own chips, is a good example. Like Blumenthal,

Turner is bearded and balding, and while he's less outgoing than his California counterpart, he's equally good-natured. Unlike Blumenthal, he's an average height, and anything but athletic. He is heavily built, with thinning reddish hair that he has trimmed once a year or so, and a pronounced limp, which is the result of a boyhood bout with polio (a reporter for *Omni* magazine once inexplicably described him as resembling a marathon runner; the description is so absurd that Turner taped it to the door of his office).

Turner splits his time about equally between theory and observation, with each split lasting, on average, a few years. "The thing about observing," he told me after dismissing the chip-bearing technician, "is that it's great fun and a real challenge. But it can also be kind of clerical—there's an enormous amount of data handling you have to do. So after I've done a lot of observing, I like to take a break and do some theory. On the other hand, theory can get to be a little bit like sitting in your office designing crossword

Edwin Turner   PHOTO: EILEEN HOHMUTH-LEMONICK

puzzles and then solving them. After a while, you get the feeling that you're out of touch with reality."

I asked him whether it was fair to say the CDM model is falling apart, and he said, "Yes, it is. Nearly every new observation we get upsets the apple cart, or at least requires the apples to be rearranged. Since the distance surveys started in the late 1970s, new observations have been causing problems.

"To some extent," he said, "the success of CDM has come by not considering the whole range of observations. People have disregarded some of the evidence. They would argue that this is not crazy, that they were paying attention only to the evidence that was well established, and this is a fair argument. But some of the less well-established evidence—the evidence for large-scale structure—has gotten stronger over time, and it has caused problems.

"My latest hobby horse—not that it's mine alone—is that the theorists look on one hand at the microwave background and on the other at the properties of the modern universe, and then try and get from one to the other. There has been a tendency to ignore what happens in between.

"There's another way to define the problem more crisply. If you think about it, there are few areas in astrophysics where a theory can supply all of the details of a system undergoing major change. In fact, we more or less don't understand how *anything* forms— dust, stars, galaxies—anything. It's not unreasonable to think that large-scale structure is a similarly complex process, yet we're asking a simple theory to account for the change from a radiation-dominated plasma to a gravity-dominated universe filled with galaxies. We really need hints from observations to let us know how the process has proceeded; we can't just deduce it from a theory."

Turner was not sure that the problem of how structure arose was ever close to being solved, but during the early 1970s, at least, he remembers that there was a consensus. "The field at that time had what was closest to a standard model. It was basically the model Peebles proposed: random fluctuations in the very early universe were amplified by gravity to produce the galaxies. Back then, there was a reasonable convergence of opinion. Then, once we started getting 3-D maps of the distribution of galaxies, and better limits on fluctuations in the cosmic microwave background, all of that fell apart.

"When I was in grad school at Caltech in the early 1970s, I mostly focused on what we could learn about omega. There was already lots of evidence for dark matter, but no one paid much attention. Then, in the early seventies, people like Ostriker and Peebles—Ostriker especially—started pushing for recognition of the consistency of the indications of dark matter. They insisted that it be taken seriously. The only systems where the dark halos weren't seen were binary galaxies, and I decided to take another look. The existing observations looked good to me, and I fully expected that mine would bear them out. I did the work at Palomar and at Kitt Peak. To my surprise, the data confirmed dark halos. At the time I was also involved with Rich Gott [a theorist now at Princeton who has lately been dabbling with the physics of time travel] in simulations of galactic clustering. These were aimed less at explaining how the structure formed as how you could differentiate between the structures that formed in an open universe as opposed to a closed one—again, the problem of omega.

"Even then there were problems. There was a compelling argument suggesting that we should see a clear difference between open and closed universes, but the computer stubbornly refused to show them. Either there was something wrong with the computer or there was something wrong with the theory of structure formation. Then, once the distance surveys appeared, I stopped working on large-scale structure altogether. It seemed clear that not only were the ideas wrong but also the theorists were misguided. Many in the community didn't react as strongly, but frankly I couldn't think of anything to do with the problem that had any link with reality. So I went on to other kinds of work.

"If you assume that omega is one—and I don't think it necessarily is—you'd like ideally to find a theory that accounts for that and also accounts for large-scale structure. That's what could have come out of CDM, but didn't. In a way, I think we're trying to force too many preconceptions into our theories. People have too strong a view of what omega ought to be. CDM is a funny case of an inhomogeneous consensus. Usually in science, there either is or isn't a consensus view, and if there is, there are also usually a small number of dissenters from it. But while CDM has become the standard model for many people, there has always been a sizable number of us who didn't think it worked so well. The CDM people seemed to be on the track of a right idea, and it explained some things extremely well, but it never quite took over the world. The thing about scientific theories is that no matter how many things they get right, all it takes is one thing wrong to show the theory isn't true. Still, there are people like Rich Gott and Jim Gunn who started out skeptical and were persuaded, which leads me to believe there's something to it. Like Newton's gravity, maybe CDM will turn out to be an approximation of the true theory.

"But I was never even persuaded enough by it to learn it. I don't like to learn things that seem to have a significant chance of being wrong—it confuses me. Now, over the past two or three years, there has been a growing sense of disenchantment, a turning over of the consensus, something like a stock market crash. Most recently, some of the high priests of CDM, including George Efstathiou and Nick Kaiser have recanted. It's as though the pope called together the cardinals, and they agreed to change scripture.

"There's one thing that has helped make CDM popular: it's a well-defined theory, so you can do lots of calculations. Ideas in astrophysics are sometimes jumped on not because there's any evidence for them but because you can do interesting calculations. It's like a theoretical gym where you can get a good workout. There are suddenly lots of papers, and then because there are so many papers people think it's something to take very seriously. That effect can happen quite separately from actual evidence. Ideas can go very quickly from being wild speculations to mainstream theory. The situation now is that some people are trying desperately to salvage CDM, or at least some parts of it, while others are running around throwing their hands up.

"One real problem in cosmology is that in physics, simplicity is very important. The basic laws are simple, symmetric, elegant. This fact has been a useful tool in evaluating theories. But the realization of these laws in the real world is very complex. Just consider the weather. I think it's unclear how cosmology fits in that scheme. The prevailing view is that the basic nature of the universe must be simple. But it's unclear how far that should go. All theories of large-scale structure try to be simple, but we don't know that the reality is simple. I'm not sure whether elegance is an asset in these theories—even in inflation. My intuition is that structure building might in fact be a messy process. I could make a list of twenty different processes that might be involved.

"What went in at the beginning was simple. What's come out is complicated. The ugly possibility for structure formation is in fact plausible. This is anathema in the astronomical community. But it could well be that there are a lot of simple ideas that have some truth to them, but none of which is the whole truth. If that's right, then the only way to really understand what went on is to look at the very early universe. We just need more observations."

CHAPTER THREE

WHERE THE GALAXIES ARE

When I first met John Huchra in his office in Cambridge, Massachusetts, a few months before the Mount Hopkins observing run, he was wearing his associate director's costume: a blazer, gray flannel trousers, and a necktie. A little shorter than average, he has a round face and a solid build, but, at the age of forty-two, he has no discernible excess flesh thanks to his taste for cross-country skiing, alpine mountaineering, and technical rock climbing. His hairline has receded several inches, and he wears a beard and thick glasses.

It wasn't apparent at the time just how uncomfortable he must have been dressed that way, but the next time I saw him, coming off the plane in Tucson en route to Mount Hopkins, he was clearly feeling much more relaxed. He was wearing a wide-brimmed hat, jeans, boots, a flannel shirt, and a leather jacket; he carried a battered metal briefcase and had a pack slung across his shoulder.

Huchra had arranged for a ride from the Green Valley Taxi Company to take him to observatory headquarters, where he would pick up a four-wheel-drive van and drive it up the rough road to the summit. At the time, Green Valley Taxi was the ride of choice for arriving astronomers (it has since gone out of business) because its owner was thoroughly familiar with their routine. Waiting for Huchra at the air terminal, Pete—he never offered a last name—explained that he always tried to pick up Huchra himself rather than sending an employee. "John is good people, and I like to treat him right," he said. It was clear from Pete's manner—he is an ex-marine and, in style if not in fact, an ex-football player—that he might not feel the same way about some of the stuffier academics. "I call him the Indiana Jones of astronomers." It was partly the trademark hat, he said. "That, and also the fact that he goes all around the world having adventures in strange places. Hey John," he yelled, spotting Huchra at the end of the corridor. "How's it going, pal? How's the new job?"

"Two years, ten months, twenty-one and a half days left," Huchra answered. "After that I won't have to be associate director anymore, and I'll get to do some science." Huchra was forgivably cranky. He was recovering from both a cold and a trip to the West Coast, where he'd been talking to colleagues at Caltech. He was exhausted, and remained slumped in the backseat of the taxi for the forty-five-minute ride. And things weren't going to let up for a while. After this six-night run, he had a week back in Cambridge, and then he was heading to Arecibo, Puerto Rico, for a stint on the giant radio telescope nestled in a naturally bowl-shaped valley.

"I tend to work on about twenty projects at once," he said. The one at Arecibo was an attempt to measure the precise distances to far-off galaxies using the natural radio waves they broadcast as a guide—a complement to his visible-light distance surveys. "The problem with doing radio observations is that you can do them whether it's night or day, rain or shine. So your schedule is determined not by when it's dark, but by when the objects you're interested in are in the sky. The last time I went there the schedule was fourteen hours on, two off, four on, four off, for a week. Then

John Huchra

PHOTO: EILEEN HOHMUTH-LEMONICK

the guy I was working with got sick, so I had to do all of it myself. I was a zombie."

It isn't that Huchra, and astronomers in general, are gluttons for punishment (although his colleagues probably believe that he is). It's that time on big telescopes, radio or otherwise, is severely limited. In decades past, the senior observers at a major telescope such as the two-hundred-inch at Mount Palomar might have twenty-five moonless nights of observing time a year—an impossibly large number today, even though there are more big telescopes now. The problem is that there are many more people who want to observe. "That's good and bad," said Huchra. "It's good to get a lot of new people into the field, since that's what brings in new ideas. But it's hard to get things done." If you have to stay up for twenty-four hours at Arecibo, you do it, because you might not get another chance on that particular telescope for months. Huchra gets so much time because his work can often be done on smaller telescopes, which aren't in as much demand; because he is on the staff of the Smithsonian, which owns the multiple mirror telescope (MMT) and several other telescopes; and because he collaborates with lots of other astronomers—as in the Arecibo project—taking advantage of their assigned time as well as his own.

The Mount Hopkins run would be a comparative piece of cake. Huchra was observing in visible light, so all work would have to stop at dawn, or long before, if clouds rolled in. He and Margaret Geller, with eight wedge-shaped slices of the universe under their belts and one Great Wall, had decided it was time to move on to a new survey strategy. They had cast their nets wide; now it was time to go deep. They were now going to map the positions of galaxies in a one-degree-by-one-hundred-degree patch of sky— that is, twice as wide as the full Moon in one direction and four hundred times as wide in the other. This new effort had been nicknamed the Century Survey, after the hundred-degree dimension, and although it would take in a region of sky they had sliced into before, it would include galaxies ten times fainter, and thus in most cases much farther away than those in the original study. The map he and Geller would produce from the Century Survey would tell them whether bubbles, voids, and Great Walls were merely a local phenomenon or were a universal feature. And it might also uncover the existence of structure on even larger scales than they expected—super Great Walls perhaps, or something even more surprising.

The reason Geller and Huchra were scanning yet another slice of sky, and why they had chosen slices rather than cones or pyramids or any other shape was made clear in a demonstration Margaret Geller gave one night at a public lecture in Princeton a few months after the observing run on Mount Hopkins. Geller is a fast-talking, intense, bright woman (Huchra, who speaks a hair slower than average, probably gets in three words to her ten), and she does not easily forget an insult. "I was treated very badly here," she told me just before the lecture. "I was the second woman to be admitted to the graduate program in physics, and a number of the faculty members made it very clear they didn't approve of me."

If she had not forgiven Princeton, though, the physicists and astronomers there were clearly trying to show her that she had long since earned their respect. They had scheduled two public lectures for her, one at the Princeton Plasma Physics Lab, where physicists are trying to tame the power of the Sun and the hydrogen bomb to generate energy and the other at a large auditorium at the Woodrow Wilson School for Public Affairs—the same room where the 1984 conference on dark matter that so depressed George Blumenthal had been held. But by far the most important talk was the third, at the weekly physics colloquium held at Jadwin Hall. "You know," she said, "this is the first time I've been asked to come back and speak to the physics department in that whole time." She was introduced to her assembled colleagues by

Margaret Geller   PHOTO: EILEEN HOHMUTH-LEMONICK

James Peebles. "I remember thinking when I saw the original survey," he said, "that I would have advised doing a less dense sampling. Luckily, she didn't ask me." It was particularly important to her that Peebles was gracious: he had been her thesis adviser and they did not get along at all.

The physics lecture went well; Geller presented her data straightforwardly, discussed its implications for the cold dark matter model of the universe ("In my view, there is really no model that satisfies all these observational constraints," she concluded), and fielded the traditional flurry of tough questions with total confidence. But it was at the public lecture—where, to her delight, Robert Dicke showed up in the audience—that she made it clear how well she can hold the attention of a nontechnical audience. The university had set up a series of talks on the universe. (Among the other speakers in the series that year had been Fang Lizhi, the Chinese astrophysicist who hid out in the U.S. embassy for a year after the Tiananmen Square massacre, a target of the Chinese government's anger for his dissident views.)

Geller's explanation of the CfA survey strategy was startlingly simple. "Suppose you were an alien making a survey of Earth, and you had only enough time to map the equivalent of the area of Rhode Island," she asked the audience (she had already told them that the amount of the universe surveyed so far is equivalent to one Rhode Islandth of the Earth). "What would you do? If you chose a patch of ocean, which you probably would, statistically, you'd conclude that the planet was covered with ocean. If you chose a patch in the Rocky Mountains, you'd have to assume it was covered with mountains. If you chose the Sahara, you'd be fooled again. How can you choose your survey so that it gives you the best possible sample of the planet? Well . . ." She held up a map of the world and large scissors and had someone take one side of the map while she held the other. Then she began cutting the map in half, horizontally. When she was finished, she let the bottom half drop to the floor, and then began cutting again, barely an inch above the new bottom of the map and parallel to the original cut.

All at once the audience gasped, as they collectively understood her point. She finished cutting and held up the result: a thin slice of Earth that took in lots of ocean, a little bit of mountain, some desert, and various bits of other terrain—an excellent representative sample of the overall makeup of the Earth's surface. She did not even have to say that the wide, narrow swath of the universe embodied in the original Slice of the Universe and the seven subsequent slices represented exactly the same sort of strategy.

Geller's clear sense of how to attack the problem of figuring out where the galaxies are is why she is the strategist of the team. "John and I decide together what the goals of the project are," she told me after the lecture. "I figure out what predictions can be tested by the surveys and thus what we should measure. John goes out and makes the measurements, and I analyze them."

During her lecture, where she talked as much about the history of mapmaking as about astronomy, she pointed out that cartographers inevitably have prejudices about what the world will look like before they go out and learn the facts. She and Huchra, she admitted, had been no exception. "We didn't expect to find any structure," she said. "We were so sure there was nothing interesting in the data that we didn't push Valerie de Lapparent to plot up the data as it came in. John and I even worried that the project wasn't good enough to make into an acceptable thesis for her. The data weren't plotted until the fall of 1985, six months after the observations, and it was then that the bubbles became evident."

It wasn't that the astronomers thought the universe was literally structureless at large scales, only that whatever structure did exist must be a fluke, like a New York taxi driver who is cautious and courteous. They knew before they started that there was at least one enormous void in space, a place substantially without galaxies, like the empty bubbles they would find later. It was this giant chunk of emptiness, in the constellation Boötes, that motivated them to do the survey in the first place.

The Great Void in Boötes, unlike the Great Wall, was found purely by accident, the result of a different sort of survey altogether. One of the astronomers who found it, in fact, wasn't even much interested in galaxies. Bob Kirshner is on the same floor as Huchra and Geller at the CfA in Cambridge, and he started out life as an expert on supernovas. He remains one: Kirshner was deeply involved in the study of Supernova 1987A, the exploding star that made the cover of *Time*. He still holds a supernova party every year to commemorate the event, which is in perfect keeping with his personality. Kirshner is the closest thing astronomy has to David Letterman. He even looks a little like Letterman: same sort of slightly freckled, rounded face and similar hair. And like Letterman, he always looks as though he's barely holding back a grin. When he is sitting in the audience at a colloquium or a seminar, the speaker has to watch out. Kirshner is apt to zero in on the absurd even in cosmology and usually can't restrain himself from commenting.

"When I finished grad school at Caltech," he told me one leaden Cambridge afternoon in midwinter, "I really wanted to stay with

Robert Kirshner   PHOTO: EILEEN HOHMUTH-LEMONICK

supernovas. But nobody will give you telescope time to study them, because you can't predict when they'll go off." So he hooked up with Augustus Oemler, a Caltech classmate who's now the head of the astronomy department at Yale, and with another astronomer, Paul Schechter (now at MIT), on an observing project at Kitt Peak. Oemler and Schechter were trying to calculate the inherent brightnesses of faint galaxies, and Kirshner figured that as they looked at hundreds of galaxies, one by one, they would also stumble across supernovas.

As they proceeded, though, said Kirshner, "we noticed something strange. The density of galaxies was different in different parts of our sample. We were focusing on the average properties of individual galaxies, but out of the corner of our eye we saw this other thing. So we said, 'let's go fainter, and see what's happening.' When we did look deeper, we took fields separated by as much as thirty degrees—that's pretty far apart—and we were confident they would be independent. But when we plotted them up . . ." He began shuffling through piles of paper. "Can I find the plots?" he wondered. "I'll bet I could, given a month. Well, anyway, there were three fields in the direction of the constellation Boötes where we could see big concentrations of galaxies nearby and big concentrations far away, but none in between. We realized something was very wrong, because the theory said the universe should be smooth at that distance. This was in 1979, 1980. Before we published the results, we went on to do the next obvious thing, so that when a lot of our friends said 'Are you sure about this?' we could say we were. What we did was to look at a lot of little fields in between the widely separated ones. We kept sticking needles into this pumpkin, and it looked hollow, like . . . well, kind of like a pumpkin. We didn't quite believe it either, but it was there, and we got a lot of publicity. This is my personal favorite." He pulled a yellowing *National Enquirer* from a shelf and opened to the middle. The headline read: DID A REAL-LIFE STAR WARS CREATE HUGE HOLE IN SPACE?

Expecting that the Boötes void was a glaring exception to the uniformity of space, Geller and Huchra had found that it was uniformity, instead, that was rare. Now they were being more careful about making predictions. Huchra, in particular, as much a pure observer as George Blumenthal is a pure theorist, has little patience with theoretical models. "I myself am very confident of what the data are in many cases," he told me on Mount Hopkins, "but the interpretation is something very different. I can tell you the recessional speed of a galaxy is twenty thousand kilometers per second, and be right to within twenty. But what that actually means? Forget it. I don't believe something's true unless you can prove it to me."

Take the flatness problem, for example. The idea that omega must be equal to one, because at .1 it is already so close, is persuasive to many astronomers. "Not to me," said Huchra. "I don't know about you, but I think a tenth is pretty far away from one. It makes a big difference to me whether I have ten thousand dollars in my bank account or a hundred thousand. I'm just not persuaded by the theoretical arguments. I trust the data." His observations, and the map that came out of them, played a crucial role in throwing cosmology into crisis, yet he refuses to take part in the game of speculation. "I'm really a cynic, professionally," he said. "I go to conferences all the time, and I don't say much. I just sit in the back and chuckle when people come to detailed conclusions based on data I know are very sketchy. It's funny, but depending on the fashion of the moment, astronomers will look at the same evidence and come to very different conclusions. It depends on the assumptions they make, which are not necessarily proven."

He gave an example. When Supernova 1987A went off in the Large Magellanic Cloud, a shapeless dwarf galaxy about 150,000 light-years away, visible only from the Southern Hemisphere, it was the first one that close to Earth since 1604—five years before Galileo made the first documented use of a telescope for astronomical observations. In order to know whether that three-and-a-half-century gap was typical or unusual, astronomers have to know how often a supernova goes off in the average galaxy (many won't be visible from Earth; the galaxy's dusty center blocks most of the stars in the Milky Way from view). "About fifteen or twenty years ago," said Huchra, "there was a controversy about that. One group said they happened every six years or so, while another said it was more like fifty. The first group was powerful, and kept the second from getting published. But the second group was right in the end."

As the taxi headed south, Huchra pointed out some landmarks in the arid countryside. "Up ahead there," he said, pointing at a mountain peak looming through the front windshield, "is Mount Hopkins." At this distance, the building housing the telescope was invisible. But off to the right, just visible over a ridge, was the much more distant spire of Kitt Peak, and through some trick of lighting the observatory on that mountain, though farther away, gleamed like a white pinhead. Kitt Peak, run by the National Optical Astronomy Observatories, is an astronomical metropolis: it houses the four-meter Mayall telescope, the McMath solar telescope, and a host of smaller devices in buildings scattered all over the mountaintop. Mount Hopkins, run by the Smithsonian Institution and the University of Arizona (which also has an installation on Kitt Peak), has only three optical telescopes of any consequence.

Also visible, just to the south of Kitt Peak, was the narrow, pointed top of Baboquivari Peak, considered in the ancient religion of the Papago Indians, who still live in the area, to be the center of the universe. From that point of reference, Geller and Huchra's map, which seems to put the observer at the focal point of the cosmos, is only a few tens of miles from being perfectly accurate.

Huchra then showed me some sights in the foreground. One was an immense, terraced hill much closer than Baboquivari and Kitt peaks, its sides perfectly straight. It was obviously an unnatural formation (looking down from the top of the mountain later on I would see that the hill was almost perfectly pentagonal in shape, and covered many square miles). "Those are the tailings from a copper mine," he said. "There are two enormous mines off in that direction. There also used to be a uranium mine around here, but I think they never found much.

"Believe it or not, we're now driving through an orchard of pecan trees," he said. The straight rows on both sides of the highway seemed to stretch for miles in all directions. "I believe they grow more pecans here than in Georgia. Do you know how they harvest them? A big machine grips the trunk and literally shakes the tree back and forth. If we're lucky, we'll see one." It seemed that he must have lived in Tucson at some point; he knew an awful lot of local lore for a New Jersey boy who had ended up working in Massachusetts. "You have to remember," he said, "that in the past fifteen years I've probably spent an integrated five years out here." (In the language of physicists and astronomers, "integration" means the adding up of small bits to make a meaningful whole.)

That five years has been marked by a number of memorable nonscientific events. "One time," said Huchra, "I was coming out here, and there had been a terrible series of storms. The Santa Cruz River overflowed its banks and washed out all the roads, including half the interstate. There was no way you could drive up the mountain, and so I figured I'd put on my backpack and hike up through Montoya Canyon. I couldn't get through, though, and in the end they took me up by helicopter. That's when I found out that the ceiling for most helicopters is about eight thousand feet. They need denser air to support them than a conventional plane would. The summit is eight thousand, six hundred. Fully loaded, the helicopter barely made it to the top—the pilot really had to gun the engine to get us up.

"Another time, a guy working for me had a heart attack up there. I called the hospital and asked them to send a helicopter. They said, 'Hold on, we'll call you back.' I couldn't believe it: here we were in an emergency, and they wanted me to wait. It turned out they wanted to get the helicopter in the air immediately, without wasting time finding out where it was going. They called me after it was airborne to get directions."

Now it was time to make a crucial stop. The taxi pulled into a shopping center and stopped in front of a supermarket. We went inside and got shopping carts. The observatory on Mount Hopkins has a well-equipped kitchen but, unlike most major observatories, no cooks. The astronomers buy their own food (except for such staples as flour, sugar, coffee, and peanut butter, which are provided for them), and everyone is assigned one shelf in the refrigerator and one in a cabinet. They do their own cooking. Huchra wasted no time buying steaks, beer, vegetables, soft drinks, and the ingredients for homemade chili. I asked him whether he was opposed to canned. "Not at all," he said, throwing a can into his cart. "You've got to realize, there's dinner and then there's what you eat at three A.M., when you've been working all night and you're desperate for something filling. One piece of advice," he added, "don't get ice cream. At altitude, the pressure in the air bubbles trapped inside is a lot higher than the air pressure outside. Ice cream tends to explode."

It was only a ten-minute ride from the market to the Whipple Observatory headquarters, an old, one-story, white-painted brick building just off the highway with offices, a tiny display area, and a gift shop selling postcards and T-shirts. Huchra checked in with the staff and then we went out to transfer luggage and four bags of groceries apiece into a beat-up GM Custom Deluxe four-wheel-drive van. "These things tend to be surplus from military bases around here," said Huchra. "We get them with a hundred thousand miles or so on them, and keep them going for another hundred thousand."

Abruptly, the road changed from two lanes of paved to one and a half of dirt, winding through dry grass and desert bushes: prickly pear cactus, ocotillo, and agave. There was no saguaro cactus, the kind that invariably appears in "Road Runner" cartoons, although it grows in the region. "Too wet around here," said Huchra, waving at the arid landscape. The van passed a sign that said, NARROW WINDING MOUNTAIN ROAD. It seemed a little late for a warning, considering the past several miles. "Just wait," said Huchra. Almost immediately, the road shrank to barely a lane, rutted and with sheer drops of hundreds of feet on one side. No guard rails most of the time—just a ridge of loose dirt maybe eight inches high. I suggested that this might not do much if the truck suddenly decided to go sideways. "Oh, it would never stop you," said Huchra. "We've had pretty good luck, though. Nobody working for the Smithsonian has ever gone off the side." Others have. "A cement truck went over once," said Huchra. "Crashed and burned. The Forest Service got it out, though: they like to keep the place looking as natural as possible. In fact, when we put up buildings we have to camouflage them as much as possible."

The wildness of the place is fine with Huchra. "I'm a hermit by nature," he said. "When I quit doing astronomy, I can imagine heading off to live by myself in the north woods. There's a lot of wildlife up here. Bobcats, mountain lions, bear, deer. The bobcats are incredibly shy. I consider myself lucky that I've gotten to see one. Our bear isn't shy at all. He's a garbage bear—he associates humans with food. I've also seen mountain lions, and I wouldn't want to run into one up close. They're probably as afraid of you as you are of them, but just barely." This was not the most reassuring news, since I realized I would have to walk down a dark road from the telescope building to the dormitory building alone if I ran out of steam in the middle of the night.

Finally the truck reached a group of buildings. This was not the summit, but the ridge where two optical telescopes are housed at opposite ends of a single building. One is the sixty-inch, where Huchra gathered most of the data for the redshift survey. He found the Great Wall with that telescope. The other is a forty-eight-inch. The dimensions refer to the diameters of the telescopes' concave mirrors, which gather faint starlight and funnel it onto photographic plates or, more commonly nowadays, electronic light detectors. Professional astronomers almost never look directly through telescopes anymore, but Huchra remembered that the forty-eight-inch did have an eyepiece somewhere. It would have been an incredible experience to look through such an enormous telescope—the biggest naked-eye telescope in common use is eight inches across, and the forty-eight-inch gathers thirty-six times as much light—but Huchra couldn't find the eyepiece.

We would come back down later for Huchra to chat with astronomers working on these smaller telescopes (he is, after all, for two years and some months more, "the boss"), read his electronic messages, and point out the Andromeda galaxy to me. But now we continued to the summit to put the food away and get settled. We found our assigned shelves, put our gear in the dorm rooms, and then sat in armchairs in the recreation room, where clouded-out astronomers can play billiards, watch the wide-screen TV, or listen to the stereo. Huchra talked about how the observing run would work and how distance surveys work in general.

If not for three fundamental facts about the physical universe, it would be impossible for astronomers to measure the distances to galaxies at all. The first fact involves the behavior of waves: when a wave-emitting object is moving toward you, the peaks of the waves are crowded together, hitting you more often than they otherwise would. When the object is moving away, they are pulled apart, hitting you less often. If the waves are waves of water, generated by a kid sister (for example) bouncing up and down in a swimming pool, the difference between an approaching kid sister and a retreating one is that the former will splash you more frequently and the latter less.

If the waves are waves of sound moving through air, the same rules apply: the sound waves generated, say, by a siren on an approaching police car splash into your ear more frequently—that is, with a higher frequency—than they would if the car were standing still. After it goes by and is moving away, the sound waves reach your ear with a lower frequency than they normally would. The human ear perceives changes in frequency of sound waves as changes in pitch, which is why the siren from a passing police car suddenly changes from a high pitch to a lower pitch; the sound drops, although the siren is broadcasting a single pitch the whole time.

If the waves are waves of electromagnetic radiation—light—traveling through empty space, the same rules apply. The eye or a piece of film or an electronic light detector, perceives the highest frequencies of visible light as violet, and on through the colors of the rainbow to the lowest frequencies, which look red. The eye stops there, but light doesn't: lower than red is infrared, then microwaves, then radio waves. Higher than violet is ultraviolet, then X-rays, then gamma rays. It's all the same stuff, and it all travels through space at the same speed. The difference in wavelength reflects a difference in energy and determines what kind of detector—eye, infrared film, radar dish—can pick it up.

When a light-emitting object is moving, its frequency appears to shift, and this frequency shift is a precise indicator of the object's speed, either toward or away from you. If a galaxy's light is twice as red as it should be—if it is redshifted by a factor of two—that means the galaxy is moving away from Earth at about 150,000 miles per second. But what does it mean to say "twice as red"? Degrees of redness are highly subjective for one thing, and for another, some galaxies are inherently redder than others.

That is where the second convenient fact comes in. Starlight—which is where the light in galaxies comes from—is made up of a mix of colors. That is evident in the aftermath of a rainstorm, when water droplets in the atmosphere dilute the sun's light into a spectrum of colors, a rainbow. The fact that the light can be smeared out is not all that helpful. But stars are made up of chemical elements—mostly hydrogen, a little helium, and tiny amounts of many other elements—and each of these elements by itself, heated to incandescence, emits not a spectrum of colors but a specific color, a specific frequency of light, a narrow line of extra brightness superimposed on the star's spectrum at a specific location. Every star has these lines, and every line always falls at the same spot in the spectrum, no matter which star you look at.

The lines are in the same spot, that is unless the star—or the galaxy made up of billions of stars—is moving toward or away from you. In that case, as the overall light from the source is violet-shifted (for an approaching galaxy) or redshifted (for a receding one), so are the lines. And while it's nearly impossible to talk about one object being twice as red as another, it's perfectly reasonable to say that the spectral line representing hydrogen alpha or oxygen has a frequency precisely half of what it would if the object were at rest. Astronomers don't use raindrops to smear out galaxy light, and they don't even use glass prisms, as Isaac Newton did in the early 1700s. Instead, they use diffraction gratings—pieces of glass on which minute, parallel lines have been scratched. They capture this diffused light onto photographic plates or electronic detectors.

The shifting of spectral lines tells astronomers precisely how fast a given galaxy is moving toward or away from Earth. Light from the Andromeda galaxy, for example, is violet- or blueshifted (both colors are at the same end of the spectrum; astronomers usually use the latter term), and the degree of shifting indicates that it is approaching the Milky Way at fifty kilometers per second. At that rate, the two galaxies should have a close encounter— maybe even a collision—in ten billion years or so.

Knowing the speed of a galaxy, even to within a couple of kilometers per hour, would be useless, however, without the third convenient fact about the universe: it is expanding. The whole thing, all of space itself, is expanding—everywhere. And the galaxies, embedded in space like raisins in a rising loaf of bread, are moving apart as a result. In the second decade of this century, an astronomer named Vesto Slipher began making measurements of the spectra—the smeared-out starlight—of spiral nebulae. This was before anyone knew the nebulae were really galaxies. Of the handful he studied, he found that nearly all were redshifted.

In the 1920s, Edwin Hubble proved the nebulae really were galaxies (he found a type of star in Andromeda whose inherent brightness he already knew, calculated how far away it must be to be as faint as it actually appeared, and thus proved it lay far beyond the boundaries of the Milky Way). That made Slipher's data all the more mysterious. If the galaxies had been gas clouds, they could plausibly be exiting the Milky Way in the aftermath of some sort of explosion. But if they, too, were full-fledged galaxies, what could possibly have caused them to flee en masse from our own? Hubble finally answered the question later in the 1920s: they weren't. He found marker stars in other nearby galaxies and compared their brightness with the galaxies' redshifts. Hubble also made educated guesses about the distances of galaxies too far away for their individual stars to be seen; he reasoned that all galaxies of a given form were about equally bright and assumed the dimmer ones were more distant. Then he compared these inferred distances with the redshifts he could observe.

The result of all these comparisons was Hubble's law, which in essence says: galaxies recede from the Milky Way at speeds proportional to their distances. If galaxy A is twice as far away as galaxy B, it will be receding twice as fast. If it is four times as far away, it will be receding four times as fast. And so on.

If it seems like a coincidence that all the galaxies have chosen ours to run away from, it is positively supernatural that they should have coordinated their speeds so carefully. The interpretation that the whole universe is expanding, though, eliminates all of the mystery (except why the universe should expand at all, of course). Again, imagine a raisin-bread universe, a space made of dough with raisin galaxies spread evenly through it. Yeast makes the dough double in size in one hour. Its height doubles, its length doubles, its depth doubles, and the distance between each pair of raisins doubles. If you were sitting on a particular raisin, you'd see that every other raisin would be twice as far away after an hour. They would all be moving away from you—and the view would look the same from every other raisin as well. It would seem to each raisin that it was in the center of an expanding universe.

Moreover, the other raisins' recessional velocity would be proportional to their distance. If raisin A started out an inch from you and raisin B started two inches away, they would be two and four inches away, respectively, after an hour. Raisin A would have gone an inch per hour, while raisin B would have gone two inches per hour. Twice as far away means twice as fast. As it is with raisins, so it is with galaxies.

With these three physical phenomena that aid the calculation of distance astronomers can now make a three-dimensional map of the cosmos. They locate a galaxy in the sky, take a spectrum, see how the lines in it are redshifted, use the redshift to calculate the speed of recession, translate that into a distance, and plot the galaxy's position. There is only one glaring inaccuracy to such a map: the scale. Because they can measure the minutest shifting of spectral lines red- or violet-ward, astronomers can say with confidence that galaxy X is flying away 30 percent faster than galaxy Y, and that it is therefore 30 percent farther away. But they can't say exactly how far away galaxy X is in the first place; in fact, they can currently get no closer than a factor of two from the right number. A given galaxy may be five hundred million light-years away, or a billion. The universe itself is somewhere between ten billion and twenty billion light-years in radius. That is like saying that Los Angeles is somewhere between four hundred and eight hundred miles from San Francisco.

The actual distance scale of the universe is a major unanswered question of astrophysics. In two of his other projects—the one at the radio telescope at Arecibo, and a search with the Hubble space telescope for variable stars in nearby galaxies—Huchra is working on just this problem. But typically, he has no opinion on what the answer will turn out to be. "I decided a few years ago just to wait."

In their role as universe-mappers, though, Huchra and Geller don't worry too much about the distance scale. They are more interested in what the structure of the cosmos looks like, in its form rather than its size. In fact, the third dimension of their map is not distance but velocity—their slices of the universe live in a place called "velocity space." Because of the direct relationship between velocity and distance, velocity space is similar enough to real space that the difference is rarely emphasized.

It is simple in theory to map galaxies, but it is not at all easy in practice. Distant galaxies are vanishingly faint, even in powerful telescopes, and when their light is spread out to reveal a spectrum they get even fainter. The familiar spectral lines, so easy to see in stars and nearby galaxies, are sometimes almost impossible to pick out from the background "noise" of scattered light from Earth's atmosphere and from dust in the Solar System. "You know," said Huchra, "ninety-eight percent of all astronomers don't know what it is to observe a galaxy. Back at Harvard, three or four out of a couple hundred astronomers have taken spectra."

Once the observing run started, it became clear that Huchra was a master: he would call up the chart of a galaxy's spectrum on a computer screen, point to a single small squiggle that looked to my eye no different from a hundred other squiggles that made up the spectrum, and say "There's hydrogen beta" or something similar. He is intimately familiar with these obscure patterns; they are like the faces of old friends. "You have to remember," he said, "that I'm . . . well, the best in the world at this. I've probably looked at about twenty thousand of these things. The next closest is Alan Dressler [one of the Samurai], and he's probably looked at three thousand. And it goes way down from there."

The most distant galaxies in the original Slice of the Universe had a recessional velocity of fifteen thousand kilometers per second, and a redshift of .04; these galaxies are only about 4 percent of the way back to the Big Bang. The edge of the Century Survey will be speeding away at forty to fifty thousand kilometers per second, which produces a redshift of .15, about 15 percent of the way back. "We already have a number of chunks of the Century Survey done," said Huchra. "Now we're working on filling in some of the intervening pieces. We need another three hundred fifty galaxies to get the two thousand we're aiming for. If the weather's good, we might be able to do it in six nights. Eventually, we want to add up all our slices and future Century Surveys to get a disk-shaped survey—thick, like a hockey puck rather than like a compact disc."

The way Huchra and Geller locate the galaxies they will study for the Century Survey is to take photographic plates from the Palomar Sky Survey, which took images of the entire northern sky during the early 1950s. Then they used a machine called a photo-densitometer to digitize the pictures, to copy them electronically into a form a computer can read. "Once they're digitized," he said, "we use programs like Tony the Tiger's—that's Tony Tyson, at Bell Labs; I have nicknames for just about everyone—which automatically weeds out the sharp images of stars in the foreground from the fuzzy images of galaxies in the background. Then, because I don't trust photographic plates, we also take electronic images of selected parts of the same stretch of sky and compare them." Then they do the hard work of just taking redshifts, one after the other.

The terrible faintness of the galaxies he would be hunting on this observing run meant that Huchra could not use the sixty-inch telescope down on the ridge. He needed to gather as much of the meager light as he could to even have a chance of reading spectra. So that afternoon, about four o'clock, Huchra got into the truck and headed, not down to the ridge, but up, following a steep and curving road 150 feet higher to the true summit of Mount Hopkins and the multiple mirror telescope (MMT). I elected to walk, and I was panting by the time I got to the top. Huchra was waiting, amused. "The problem is that you're at eighty-six hundred feet, and your body just doesn't like going uphill."

The telescope building sat in the middle of the mountain's (artificially) flattened top, a plateau perhaps three hundred feet across and covered with gravel. It looked like no ordinary observatory. There was no dome, just a building more or less cubical in shape, except for the roof, which slanted slightly. On one wall the cube had what looked like two enormous barn doors covering the upper two thirds of the building. (I would find out later that there were barn doors on the roof as well). When the telescope is in operation, the doors slide aside, creating a much wider gap than the traditional slit big telescopes usually look through. The opening lets air circulate freely, bringing the temperature inside the dome into equilibrium with the air outside, and eliminating the heat-generated air currents that can make stars and galaxies look too twinkly for precision work.

The telescope is as unusual as the building that houses it. Astronomers' constant yearning for ever more light, so they can study fainter and fainter objects, ran long ago into a physical barrier. As the size of light-gathering mirrors grows, their thickness and weight grows as well. That makes them expensive to buy (telescope mirrors are made out of special glass that expands and contracts minimally with temperature changes) and expensive to transport. It also makes them tend to sag under their own weight, distorting their shape enough so that they are useless for astronomy; the sag shows up in different parts of the mirror every time the telescope moves to a different angle. It's always possible to keep adding girders and supports to keep the mirror stiff, but because the whole structure has to move, that again is very expensive.

The last very big, high-quality mirror made in this old-fashioned way is the venerable two-hundred-incher on Mount Palomar, in Southern California. It was until 1992, when the Keck telescope in Hawaii surpassed it, the biggest effective optical telescope in the world (a somewhat bigger mirror in the Caucasus Mountains of the Soviet Union has never worked properly). Modern telescope designers have experimented with a number of ways around the problem. Ermanno Borra, a professor at the University of Laval in Quebec, has suggested making mirrors out of liquid mercury that sits in a shallow but enormous rotating pan. The shiny liquid, forced toward the edges of the pan, forms a perfect parabolic curve, the kind mirror-makers strive for. Unfortunately, a liquid mirror telescope, which already exists as a small, laboratory prototype, can only point straight up. Tilt it, and the mirror spills.

There are also more useful innovations in mirror-making that have been developed, but for its time the MMT was a work of genius, and it is still a useful alternative to single big mirrors. Before we went into the control room to set up for the night's run, Huchra took me into the telescope chamber itself to show me how it worked. Instead of a single, large mirror strung up in the telescope's steel framework, there were six relatively small disks of glass, each of them seventy-two inches across and arranged like the chambers in a six-shooter. There is no barrel on this or any other large telescope; the barrel on backyard telescopes screens out stray light, but here the building itself serves that function. Open to the desert air every night, the mirrors, like those in most major telescopes, were covered with a thin layer of dust; lying right on the mirror, the dust is so out of focus that it doesn't show up at all on the final images.

The mirrors were fixed in place within the framework, but suspended above each one was a small, movable secondary mirror whose job was to bounce the starlight focused by each seventy-two-inch mirror toward whatever light detector the astronomer of the week was using. These secondaries can move closer or farther from their primaries to sharpen the focus, but they also swivel. That allows them to work in concert to combine the light from six separate primaries into a single beam. The MMT is thus six separate telescopes that work together to simulate a single, expensive telescope with a mirror 176 inches across. "Within the next few years," said Huchra, "we're going to fix the holes." What he meant was that the six mirrors were going to be scrapped in favor of a single, ultralightweight 254-inch mirror, made with a new technique by Roger Angel, of the University of Arizona.

We crossed the hall into the control room. It was about ten feet wide and thirty long, and dominated by two big consoles, each holding keyboards, switches, digital readouts, and other electronics. In total, there were fifteen video screens. Most were used to display computer code and plot data, but two of them (one on each console) showed an image of what the telescope was seeing; they are the electronic equivalent of the small finder telescopes used to point big telescopes decades ago. Huchra immediately went to work with the postdoc, Tadashi Nakajima, whom he'd be working with that night. Nakajima was studying spectra, too, but in order to understand in detail the kinds of elements found in nearby galaxies, not to determine the redshifts of distant ones. The two astronomers, master and apprentice, started setting up the control room, adjusting monitors, working on computer routines. "Hand me a scissors," said Huchra, and he began cutting and splicing wires.

At that moment, a slightly heavy, gray-haired woman came in, dressed in jeans, a sweater, and a well-worn parka. "You'd think that hike would get easier, since I do it every day," she said, "but it doesn't." Her name is Carol Heller, and she is the telescope operator. The first thing she wanted to do was see what tapes and CDs Huchra had brought. "On a long winter's night," Huchra explained, "you have to listen to music to keep from going to sleep. Your average observer will come up and listen to Bach all night, but some of us are a little more adventurous. I prefer Jefferson Airplane."

As she began to organize her own computers and control devices, Heller talked about her job. "I'm one of four people who are allowed to operate this telescope," she said. Astronomers are not among them. The MMT is an exquisitely sensitive scientific instrument that happens to weigh dozens of tons, and no one is about to hand over the controls to an inexperienced driver. "Even if you could train them all," said Heller, "and keep them up-to-date on instrument changes—which would be a formidable job—even then, they would only be using it every three or four months, so they'd be out of practice.

"Telescope operators come from all sorts of backgrounds," she said. "Training is strictly on the job. Once you can do it, though, I imagine you can go anywhere in the world they have a big telescope. I began on the ninety-inch at Kitt Peak. I had just graduated from the University of Arizona with a degree in ecology, and there were no jobs. But I was interested in astronomy, too, and I kept pestering the personnel department for a job. One finally came up, and there were about ninety applicants. Some of them weren't very knowledgeable, though. They just thought it sounded like a neat job. It didn't even occur to them that they'd have to stay up all night. You also had to be willing to climb up on the outside of the dome and push snow off the roof. In fact, instead of asking me a lot of questions, they just sent me up on the dome to see if I would freak out. I didn't.

"The other test they made me take for the job was lifting a heavy instrument and attaching it to the telescope, and crawling inside the thirty-six-inch telescope to make an adjustment. They gave me a pretty rough time, maybe because I was a woman applying for a job usually done by men."

"I did have a rough time one night after I had gotten the job, though. It was two A.M., and a storm had just come through. The wind was still blowing about forty miles an hour, and I had to go up a ladder and knock ice off the anemometer [it measures wind speed]. The rungs were covered with ice, and about halfway up, the ladder started swinging back and forth—I was just paralyzed. I might have felt better with a rope, and I had used one before, but once when I was up there somebody began opening the telescope shutter, which got caught on the rope and started pulling me off the roof. I got out of that rope pretty quickly."

It was now five o'clock, two hours before it would be dark enough to begin doing astronomy. Heller started concentrating in earnest on her preparations, and Huchra announced it was time to go down for dinner. The MMT has its own minikitchen, but the food and the full-sized kitchen were down below so we headed back downhill.

Huchra had decided tonight was the night for chili. He opened his briefcase, shoved aside some papers and a calculator, and pulled out a can of kidney beans and a jar of hot salsa. He began throwing ingredients into a pot. "Cooking is something of an avocation for me," he said, with a clear hint of pride in his voice. "When I finished grad school at Caltech I got two job offers at Palomar, where I'd done my observing. One was to be the cook. The other was to leave town and go to Kitt Peak. They were willing to pay me to get lost. The reason was that it seemed to be cloudy every time I went observing. They figured that if I took my luck over to Tucson the weather would have to get better in Southern California. Astronomers, believe it or not, can actually be competitive just like normal people. I remember riding down from the observatory with Uncle Allan when someone else was scheduled to observe, and he would be chuckling as he saw the clouds rolling in." Uncle Allan is Allan Sandage, the direct intellectual heir of Edwin Hubble. "Anyway, I'm not superstitious about the weather," he said. "Just pessimistic. I predict a heavy snowstorm within a day or two."

By seven o'clock we were back up the mountain, and the night's work began. The first task was to focus and calibrate the spectrometer, a maneuver analogous to fiddling with the knob on a bathroom scale so that it actually reads zero when no one is standing on it. In this case, the team focused on a star whose spectrum is well known so they would have something to compare their galactic spectra with. Huchra reached across his console and opened a file box filled with five-by-seven file cards—they record information on all the objects he might be interested in looking at—and he and Heller agreed on a likely star. They pointed the telescope at it, took its spectrum, and committed the spectrum to the computer's memory.

"Okay," he said, "we're in fat city. Let's go to ten sixty-eight." It's full name is NGC 1068, and it is a spiral galaxy on Nakajima's list. Within seconds a pinwheel of light appeared, tilted almost face-on, on the two video screens that matched the telescope's field of view. "Carol, can you check the stack?" Huchra asked. The image was a little blurry, and he suspected that the secondary mirrors were subtly out of alignment. They were not all focusing light on the same point, and the six star images were not stacked perfectly on top of one another. She guided the telescope off the galaxy and onto a nearby star and flicked a switch. A box appeared, surrounding the star, and suddenly it split into six boxes and six stars, which zoomed outward to form the points of a hexagon. Then the boxes zoomed in again, passing through each other and changing places with their opposite numbers like square dancers executing a well-rehearsed maneuver. Then they collapsed back into a single box, and the star inside looked significantly sharper than it had before. What Heller had done was to deliberately pull the mirrors completely out of alignment and let the computer put them back in a more precise configuration. The process would be repeated several times through the night, as the secondary mirrors crept slightly out of position.

Satisfied, Huchra and Nakajima settled into a routine: they asked Heller to find a particular galaxy listed on one of the five-by-seven cards. Then they fine-tuned the positioning themselves, moving the image a fraction of an inch up, then over until it was positioned right on top of a fine black dot at the top center of the screen. The dot marked the position of one of two slits that admits light into the spectrometer (there was a similar dot in the bottom center of the screen). The astronomers would take two spectra of each galaxy, one through each slit, to make sure they were getting an accurate reading. They instructed the computer to have the spectrometer take data for five minutes, or three, or ten, depending on how bright the galaxy was and thus how long it took to get enough light from it.

A small glowing display flashed numbers continuously—the numbers constantly changed, but they stayed pretty close together while the telescope was pointed at a given galaxy. First 517, for example, then 592, then 554. The spectrometer was counting photons, the individual packets of energy that are the smallest possible bits of light, striking the detector every second. After traveling millions of light-years or more, these few bits of light, far too faint for the human eye to detect, were smashing into a bit of silicon not much bigger than a postage stamp, while just a few feet away their traveling companions for the past several million years smashed, unseen, into rocks and GM vans.

Periodically, Huchra and Nakajima would call up a plot of the spectrum in progress, oohing and aahing over particularly interesting features. Sometimes they would realize that one of their reference catalogs was wrong in how it classified a galaxy or what redshift it listed or how bright it said the galaxy should appear. Then, when both slits had peered at the galaxy, the astronomers focused their spectrometer on a reference lamp inside the dome and took an exposure of that. They know precisely what the spectrum of the reference lamp is (it's a bulb filled with helium, neon, and argon gas), and that alerts them to any misadjustments in the spectrometer. And then they called out another galaxy, and Carol Heller swung the telescope around. It made the whole building shake.

In fact, the building was doing more than shaking. At one point, during an especially long exposure, I went outside, partly to look at the night sky, invisible from the windowless control room, and partly to get my courage up for walking down through the mountain-lion- and bear-infested darkness to the dorm. Immediately I realized that something was wrong. As my eyes adjusted to the pitch-blackness, I could dimly make out a fence in front of me. There had been no fence in front of the door when we entered. I heard the flag flapping in the wind to my right; it should have been slightly to the left. There was only one door into the building, and I was beginning to worry about things more supernatural than mountain lions. Then I wondered: could it possibly be that instead of the telescope swinging around, it was the entire building that moved? I went over to the metal stairs I had just descended and felt around the bottom. They ended about six inches above the gravel and didn't touch the ground at all. Suddenly a bell rang, and I jumped onto the stairs. Sure enough, the building began to rotate like a carousel, the lights of Tucson moving slowly from right to left.

By one o'clock the astronomers were just hitting their stride, but

I was worn out and I started walking back down, talking to myself in a loud voice in case the mountain lions were hungry. Huchra looked incredulous when I announced I was giving up. He couldn't imagine anyone being tired so early in the evening. (I would, over the next year of visits to observatories all over the Western Hemisphere, earn a reputation among astronomers as someone without much stamina.)

I didn't see him again until the next night; they had finished up around six A.M., and Nakajima was already on the way to the airport to catch a cheap midnight flight. Huchra was dubious about the night's observing. There were high, thin cirrus clouds overhead, too wispy for most people to notice, but potentially murderous for an astronomer searching for faint galaxies. "I told you, this is going to turn into a major snowstorm," he said at dinner.

As he cooked a couple of steaks for himself, Huchra told me that his present position in life had come about largely as a result of a series of accidents. "My father was a freight conductor," he said, "and we didn't exactly have a lot of money. That's unusual right there for an astronomer—the vast majority of my colleagues come from families that are either rich or at least very upper middle class. You'll probably find that I'm one of the most conservative astronomers around—financially, that is; it doesn't mean I'm a Republican or anything."

He went to MIT on a scholarship and was, he claimed, just an average student. "Well," he said, "maybe in the top twenty-five percent. I'm convinced the reason I got into graduate school at Caltech was that someone got my grade in a physics course mixed up with another guy's. I expected a C and got an A; he expected an A and got a C. Maybe I also got some tremendous recommendation from a professor. Anyway, I packed everything I owned into my car—it fit easily—and drove to Pasadena. The first month there was the loneliest of my life. I was the dumbest one in my class—no, actually the second dumbest. The other guy became a photographer specializing in nude models, last I heard."

The second piece of luck came when he was about to graduate. "I had gotten a job offer as a postdoc at Mount Stromlo Observatory in Australia," he said, "and I was packing to go when I got a telegram. Parliament had just taken a vote of no confidence, and the government had fallen. All government jobs were frozen. I could still come, of course, but they wouldn't pay me. So I withheld my thesis, which meant I got to stay at Caltech another year. That was great—I had a chance to dot the *i*'s and cross the *t*'s on the thesis, and I got to dabble in all sorts of things. I worked on the geology of asteroids and meteors, for example."

At the end of that year, Huchra applied for two positions, one at Kitt Peak, where they usually hired observers like him, and one at the Harvard observatory, where they usually took theorists. He didn't get the first job ("They hired a theorist, whom I know you've never heard of") but to his surprise he did get the second.

After two years at Harvard, two jobs again came up. One was as an assistant professor at Harvard, and one was a term appointment at the observatory. No one turns down a chance to join the Harvard faculty, but Huchra did. "I knew that the Smithsonian owned all the telescopes," he said. The Smithsonian job was only for a four-year term, but while he was there an opening came up for a tenured infrared astronomer. He applied, figuring that he had done a little bit of infrared observing, but the Smithsonian didn't see it that way. They asked him to withdraw his application. He said no, and in the end they created a new position for him. "As a result," he said, "I got tenure before a lot of people who had been there longer. I went from being low man on the ladder to a more secure position than anyone else my age."

It was now nearly dark, and we headed once more for the telescope. The clouds were worrisome, but to keep the early evening from being a total loss, Huchra had tonight's telescope operator, John McAfee, swing the MMT over to focus on a much brighter object: a globular cluster orbiting M31, the Andromeda galaxy. Most full-sized galaxies, as far as we know, have hundreds of these spherical blobs, each containing a million or so stars, orbiting around them. The globular clusters are not confined to the galaxies' main disk, though; they travel far above and far below the galactic plane.

The globular cluster was big and bright in the video screen despite the clouds. "I'm using this to check my redshift calibration," he said, "because we know its recessional velocity precisely. But I'm also interested in globular clusters for themselves. As long as seeing is bad and the sky is bright, I'll bag a few of these." The cluster looked almost as big by itself as the entire galaxies we looked at the night before had. I asked Huchra how big Andromeda would appear if the MMT could capture the whole thing at the same magnification. He calculated a moment, then held his hands wide apart. "Sixty times as big as this," he said. "About as big as the entire observatory."

The thought of Andromeda reminded me of something Huchra had begun to explain back in Cambridge. It was his favorite example of how theorists get on bandwagons without having all of the facts in their possession. I hadn't quite followed him then, so he explained it to me at length now. It is useful to know how fast the Milky Way is moving through the universe—not how it moves with the general expansion, but what additional motion, caused by the gravity of nearby masses, is overlaid on that expansion—the so-called peculiar motions that Vera Rubin had begun to uncover and the Seven Samurai had used to find the Great Attractor. But to know how fast the Milky Way is moving, astronomers need something to compare the motion to, something that itself can be reasonably considered to be at rest. The microwave background radiation, the glowing echo from the Big Bang is perfect, for although it moves at the speed of light, it is moving in all directions at once. Averaged over the universe, it is perfectly still.

Astronomers can measure Earth's speed in relation to the microwave background in the same way they measure speed in relation to galaxies: despite its general uniformity, the microwave background is just a little bit higher in frequency—it is violet-shifted—

in the direction of Hydra, and it's redshifted at the opposite point in the sky. The degree of shifting in each direction matches the other perfectly, corresponding to a speed of about three hundred kilometers a second. But that's just a raw number, a final velocity that's a combination of Earth's motion around the Sun, the Sun's motion as it orbits the core of the Milky Way, and the Milky Way's motion within the Local Group of galaxies.

When all these factors are subtracted out, the underlying motion through the stationary microwave background radiation is more like six hundred kilometers per second, a respectable clip. Here's the problem, said Huchra. "It's easy to measure the radial velocity between the Milky Way and Andromeda, our motion toward each other. But we can't measure our transverse velocity, our motion at right angles to Andromeda." Even if we were passing at thousands of kilometers per hour, that would be too slow for us to see Andromeda creep across our sky. "It's almost inconceivable that there isn't any transverse velocity," said Huchra, "yet because it can't be measured, the theorists simply assume it's zero." The transverse velocity might in fact be close to zero, he said but it might equally as well be large enough to cancel out part of the six hundred kilometers per second, or large enough in the opposite direction to make it a gross underestimate. "There is a handful of us who actually make these measurements," said Huchra, "and then there are the hundreds of theorists out there who don't think about the limitations."

Even observers can make faulty assumptions. In 1962, said Huchra, Allan Sandage wrote a classic paper in which he argued that you can measure whether there is enough mass to close the universe simply by counting the most distant galaxies with big telescopes. If the universe is closed in an Einsteinian sense, the narrowing curvature of space means you should see fewer galaxies far away, on average, than you do nearby. If the universe is open, you should see more far away. And if it's flat, the average density should remain the same.

But the argument depended on the assumption that galaxies

were about as bright in the past as they are now (the light now reaching Earth from these farthest of galaxies began its journey billions of years ago, after all). "In the 1970s," Huchra said, "that picture was trashed, because people realized that galaxies evolve. And it's essentially impossible to know whether they have gotten brighter or dimmer over time. With rare exceptions, all global tests of cosmology have been thrown out, and that's why when someone claims to have found one, the rest of us chuckle.

"So when programs like Sandage's died, people tried to determine these cosmological numbers locally. And the assumption is that what you find to be true locally is true globally. That is not a well-tested assumption. Say our own region of space is overdense or underdense, compared to the universe-wide average, by a factor of only two or three. That would still lead to profound errors in our estimates of the total mass density of the universe. That's what's so unsettling about what we've learned in the redshift survey. It shows that the assumption of homogeneity is wrong. Two decades ago, we all would have said that once you get out to a distance of a few thousand kilometers per second [Huchra was still thinking in velocity space] the cosmos is uniform. Now we know that it isn't true even at ten thousand. It's an interesting philosophical problem: the more we know the less we can be sure about what we know.

"Yet people hold on to their assumptions, even though history has tended to deflate them, because without assumptions you can't say anything about what you think is going on. My own answer is not to say anything, except 'give the observers time.' Then you can make statements that mean something. It's fortunate that I take pleasure in the hunt, the chase. The actual killing is not so much fun. Finding a redshift for a galaxy gives me a little bit of immediate pleasure. Unlike those who want big answers, I can solve problems every night. It's instant gratification.

"I'll give you an example. Another one of the projects I'm involved in is an international team that's looking for optical counterparts of the X-ray sources found by the ROSAT satellite [a

European X-ray observatory that went up in 1990]. We expect a lot of the sources to be clusters of galaxies, since the hot gas in clusters gives off a lot of X-rays. One of the first ROSAT clusters I've located is at a redshift of point one. Now George Abell is supposed to have found all the galactic clusters at a redshift of point one, but he didn't find this one. Who knows how many more there are? Give us five years, and we'll have a much better idea about how much clustering there was a few billion years ago.

"The difference between observers and theorists is that we observers aren't supposed to make mistakes. In my younger days, at Caltech, there was a series of theoretical seminars run by Wal Sargent, Jim Gunn, and Peter Goldreich. Every term there was a different topic and a tremendous amount of reading. They were called the Russian roulette seminars because you walked in, and everyone threw names in a hat—even the distinguished visitors. Whoever's name was pulled would have to give a talk. One day, Fred Hoyle was there, and we got into a discussion over a beer on the philosophy of science. Hoyle said, 'A good theorist is right five percent of the time. If an observer is wrong as much as five percent of the time, he should quit.' I would add that theorists who are right one hundred percent of the time are no good either—they aren't taking chances, aren't adding anything to the field.

"The best observers are the ones who go into a robot-like state, a trance. When I'm observing, I'm not thinking at all about pushing back the frontiers of knowledge. I'm thinking about how long I'm going to integrate on this galaxy, what galaxy I should be moving to next, what I'll switch to if some thin clouds come over and I can't do faint galaxies for a while. That's why theorists usually don't make good observers. They think too much. I'm good because I can stop thinking.

"There are a few people who cross boundaries, but most fit into the class of theorists, observers, instrument-makers or phenomenologists—people like Margaret Geller, who work at the boundary between theory and observation, trying to reconcile them."

One limitation theorists never have to worry about is weather, and Huchra was having a run of his graduate-school luck. The clouds kept rolling in, and in the end he would only observe a couple of globular clusters and get redshifts for seven galaxies that night. The next night was even worse, with an overcast so bad that even Huchra didn't think it was worth going up to the telescope. He and McAfee stayed down in the recreation room, playing billiards and watching old movies, and Huchra managed to get some paperwork done. The snowstorm he had forecast never materialized, but he would only get in one solid night of observing during the six-day run.

I asked him between billiard shots whether there wasn't an easier way to do the redshift survey than traveling out to a lonely mountaintop with only a chance that good observing would be possible. Princeton University and the University of Chicago had just announced their intention to build a telescope dedicated to the biggest redshift survey ever contemplated. It will be designed to permit remote observing by astronomers thousands of miles away. The Chicago-Princeton telescope is supposed to go into operation in 1997, and will eventually measure a million redshifts. Why not just wait?

For one thing, Huchra isn't completely comfortable with automation. "I'm a real old-timer at this game," he said. "There's been a lot of effort to make things easier. We now observe in warm rooms rather than freezing domes, and we guide the telescope with TV cameras that can see fainter than the human eye. But some things are more complex than they need to be. Electronics make it easier, but the catch comes when something goes wrong. Electronic systems are a lot harder to debug than mechanical ones."

The Chicago-Princeton telescope is still an interesting project, he agreed, but he pointed out that by the time they're up and running, the Harvard-Smithsonian survey will have measured a hundred thousand redshifts. "I'm not sure," said Huchra, "that you need much more than that to start getting a pretty good idea of what the universe really looks like on a very large scale."

CHAPTER FOUR

THE EDGE OF THE UNIVERSE

T he Andromeda galaxy, two million light-years away from Earth, is by far the most distant object you can see with your naked eye. For anything farther, you need a telescope or binoculars. Even a modest backyard telescope can, in a dark site like rural Arizona, bring hundreds of galaxies into view. The light-gathering mirror in a four-inch telescope—a telescope small enough to fit into a bulky briefcase, and that costs a few hundred dollars—funnels the photons that fall on more than twelve square inches of mirror into your eye all at once and thus reveals objects that are much too faint for the eye alone to see. In general, the fainter a galaxy is, the farther away it is. And because bigger and bigger telescopes bring ever fainter and more distant galaxies into view, the largest telescopes see most deeply into the universe. The multiple mirror telescope, with its 176-inch-equivalent, light-gathering surface can see hundreds of millions of galaxies in the skies of the Northern Hemisphere, some of them lying billions of light-years from Earth. It can also detect quasars, which are even more distant on average than galaxies, because they are intrinsically brighter. The MMT and other telescopes in its class can see objects on the order of ten or fifteen billion light-years away—objects so distant that the light arriving from them today began its journey long before Earth or the Sun existed.

The quasars are commonly thought of as the farthest objects humans can see, and that is true if you're talking about visible light. The most distant quasars are so far away that their light is redshifted almost fivefold by the expansion of the universe. At a redshift of five, these quasars are about 80 percent of the way back to the Big Bang; the light that is reaching Earth today left them when the universe was only 20 percent of its present age.

But if you include all the forms of light, the quasars are not the farthest. Switch from visible light to radio waves, especially the short-wavelength radio waves known as microwaves (the same microwaves that can reheat coffee or, under the name of radar, detect ships and aircraft), and you can literally see to the edge of the universe. It is an edge in time, not space; it marks the time, a few hundred thousand years after the Big Bang, when radiation was first able to move freely about the universe, when the temperature of matter fell below the level, about five thousand degrees, where light could shine through it. The cosmic microwave background is that radiation, about as bright as the Sun when it was generated, but redshifted a thousandfold by the expansion of the universe. The cosmic microwave background radiation, emitted at a redshift of a thousand, began its journey to Earth when the universe was less than .01 percent of its present age. It is the very oldest light we can see; it comes from 99.9 percent of the way back to the beginning.

If the cosmic microwave background is the inner edge of the universe, then the Milky Way is right at the outer edge, in the same sense. Light that's reaching us from Andromeda is already two million years old; it's taken that long to get here. Light from more distant galaxies is even older. Light from the Milky Way is the most up-to-date light there is, and carries information from the farthest evolved part of the universe that can ever be accessible to us.

It's a little more complicated to describe the edge of the universe in space rather than time. Basically, there isn't any. If the universe is closed, then it's analogous to a beach ball. The ball's two-dimensional surface curves in the third dimension. Run your hand around it, and while it's finite in size, you'll never find an edge. A three-dimensional, curved universe does the same thing, although the "surface" has three dimensions and the curvature happens in an unimaginable and imperceptible fourth dimension. If the universe is open or flat, and thus infinite, it's meaningless even to speak about an edge.

There was light in the universe even before the time of the cosmic background radiation: photons of high-energy electromagnetic radiation emitted by the superheated matter of the early universe. The same thing happens in a light bulb when hot atoms in the filament spray out photons of visible light. In the early universe, though, the light couldn't travel very far before being absorbed. The reason is that at high temperatures, matter is ionized: atomic nuclei cannot hold on to their electrons, and any atoms that form are immediately ripped apart. Naked nuclei and electrons are electrically charged (the charges are opposite and neutralize each other when the nuclei and electrons form into atoms), and electrically charged particles are voracious absorbers and emitters of photons. For the first several hundred thousand years after the Big Bang, light was constantly emitted, then absorbed, then emitted again, then absorbed. The universe was a thick soup of matter permeated by a diffuse fog of light brighter than the Sun.

When the temperature of the universe dropped below five thousand degrees, though, it was finally cool enough for nuclei of hydrogen and helium and deuterium and lithium to hold on to their electrons. Because the nuclei and electrons could combine for one last time without being torn apart, this event is called recombination by astronomers and physicists. It's alternatively called decoupling, because light and matter, until then intimately tied together in the cycle of emission and reabsorption, became relatively independent of each other. The light that could go nowhere until decoupling was finally able to shine freely. It has been spreading through space ever since. Because all the light from earlier times was scrambled by matter, the electromagnetic radiation that escaped at the time of decoupling, and which has been shining more or less unimpeded since then, is the earliest thing we can ever hope to see—light from the edge of the visible universe.

This makes it a crucial probe of cold dark matter and every other theory of how structure formed. The cosmic microwave background should have escaped from its confinement by matter bearing some imprint of that confinement—in general, a little warmer where matter was a little denser, a little cooler where it was sparse—and thus of the molehills that grew to be galaxies and clusters. As the radiation has spread through the universe since decoupling, it should have been modified slightly—again, showing patches of warm and cool—as it passed through the structures that were forming soon after decoupling. It should have encountered and remembered its encounters with the blobs of collapsing gas that turned into galaxies. The marks it acquired in those encounters should, by virtue of their physical size and their degree of departure from the average temperature, help vindicate or vitiate any and all theories of structure formation.

The edge of the universe was first identified in 1965 by two groups of scientists working completely independently of each other. The scientists who found it weren't looking for it and didn't know what it was when they saw it; the ones who set out to look for it weren't quick enough to find it. A little more than a quarter century later, I went to see Robert Dicke, who had been the leader of that second group, in his office in Jadwin Hall on an early summer day in the middle of a thunderstorm. Not only was the room dominated by the usual shelves full of books but also by a wall of filing cabinets, by far the biggest collection of filing cabinets I had ever seen in a scientist's office. "One problem with getting old," said Dicke, "is that you've published so many papers over your career that you can't find the one you want." Dicke looked older than his seventy-five years; he walked slowly, hunched over, and his right hand trembled slightly. He looks the part of the classic movie scientist, with a shock of white, slightly unkempt hair, and a pair of elfin ears.

Dicke's comment about his publications is misleading. His production of scientific and technical papers is not a product simply of his age. Few physicists could have published half as many papers in a career twice as long as his. Few have worked on so many problems or done so much fundamental research, both theoretical and applied, as he has. At the MIT radiation laboratory during World War II, Dicke was among the handful of physicists and engineers who invented radar. During the 1950s, he did much of the pioneering work that led to the invention of the laser. He coauthored a theory of gravitation that attempted to improve on Einstein's general relativity (it is probably wrong, but considered brilliant by other physicists nonetheless). His recollections were peppered, as no other physicist's could be, with seminal, personal references to major themes in modern cosmology. "When I invented the anthropic principle . . . ," he would say, or "When I first began to worry about the flatness problem . . ." It was a lecture by Dicke that got Alan Guth on the road to the inflationary universe.

I noticed a photograph on his bulletin board that showed some sort of scientific instrument sitting on what was clearly the Moon, with the unmistakable outline of an astronaut's footprint in the lunar dust beside it. "Oh, yes," he said. "That's one of the three-corner reflectors like they left on the Moon." The device is a suitcase-sized box embedded on one side with cells, each containing what looks like the inside corner of a box made of mirrors. A light beam entering one of the cells will always bounce out again back to where it came from. By shining a laser beam at the box from Earth and measuring the round-trip travel time, scientists

Robert Dicke   PHOTO: EILEEN HOHMUTH-LEMONICK

have gauged the distance to the Moon within fractions of an inch.

In this instance, again, Dicke's contribution was fundamental. "My group used to meet one night a week," he said, "all of us who were interested in gravity. We thought of putting one of these three-corner reflectors on a satellite orbiting the Earth, so we could see how the local gravitational field in different parts of the planet affected the satellite's orbit. Then someone showed up one night with a reflector embedded in a rubber ball that was weighted so it would always land facing up. He said, 'Why don't we throw one of these onto the Moon when we go by?' This was long before we actually went to the Moon, but by the time we did NASA had taken up the idea. You know, those reflectors are the only experiments on the Moon that are still working."

In the early 1960s, Dicke came up with the idea he is best known for, the idea that the universe might have an edge. In this case he wasn't the first to have the idea, but it was original nonetheless. The Russian emigre physicist George Gamow and two colleagues, Ralph Alpher and Robert Herman, had actually considered the idea of a superheated, compact early universe two decades earlier, and written about it, but Dicke did not, at least consciously, know about that work. "I remember Gamow giving a talk here, but I don't remember him mentioning any detectable radiation left over from the early universe. But he might have, and I admit we should have known about his work anyway. We were deficient in not reading the literature very carefully."

Deficient he might have been, but considering his track record no one believes Dicke didn't come up with the idea on his own. "I was thinking about a hot early universe," Dicke told me, "because I was impressed with the fact that the stars that formed first in the universe had very few metals in them." In astronomical terms, metals are not just iron and aluminum and copper, but any of the chemical elements—oxygen, carbon, nitrogen—that have atoms heavier than those of hydrogen and helium. The fact that these earliest stars had no metals implied that there were none around at the time they formed; otherwise, the metals would have been sucked in by the gravity of the stars and incorporated into them.

There is no problem understanding why later generations of stars, including the Sun and the planets that formed around it, should be full of metals. The nuclear reactions that take place in the centers of stars—essentially, they are gigantic hydrogen-bomb explosions—create metals as a by-product. When the young stars get old and die, they release these metals to interstellar space, either by exploding or by more calmly blowing their outer layers away in a puff of stellar wind. The metals are then available for the next generation of stars to form out of.

The thing that bothered Dicke about the lack of metals in early stars was that he was working at the time (and still was when I spoke with him) on the notion that the universe oscillates, expanding for a while as it is doing now, then contracting, then expanding again. "The problem was," he recalled, "that all the metals that should have been made inside stars in the previous cycle had somehow disappeared before the current cycle began. The only thing I could think of was that the universe had reached a temperature high enough to dissociate atoms into protons, electrons, and their antiparticles." The temperature Dicke had in mind was about ten billion degrees. "Then," he said, "as the universe expanded and cooled, the particles would be able to combine once again to form neutral hydrogen, out of which the first stars formed." Dicke knew that while they were still very hot, these particles should have produced showers of photons.

"I had been thinking of all this in a very theoretical way," said Dicke, "when I suddenly realized, Hey, this radiation should still be there to be seen." It would no longer have much energy, since in the ensuing billions of years the universe had expanded enormously, and the wavelength of any radiation left over from the beginning would have stretched accordingly. Dicke figured it should now be detectable, if it existed, at microwave wavelengths—in the same range as the radar waves he had worked with during the war. "That made it seem like an easy experiment," he said, "since we had already built radiometers at the radiation lab at MIT."

In fact, the experiment had already been done in the 1940s. "We were having trouble with our K-band radar, and we thought the problem might be that water vapor in the atmosphere might be absorbing the radar waves. So we built four radiometers to test it, and while we were at it we pointed them at the sky, both in Massachusetts and in Florida." Dicke thought there might be microwaves streaming in from space, although at the time he thought they should be coming from faraway galaxies, not the early universe. "We actually got an upper limit of twenty degrees on any background radiation—if it was there, it was cooler than that."

With a little bit of improvement, Dicke figured, his radiometers could detect radiation even cooler than that, to within a few degrees above absolute zero, where, calculations showed, the remnant radiation from the edge of the universe should be hovering. "I got a few of us together," he said. "I suggested to Jim Peebles that he look into the theoretical questions, what the expected temperature should be of cosmic background radiation. And Dave Wilkinson took charge of making the radiometer work."

When the equipment was built, Wilkinson, who now occupies an office two doors down from Dicke's (Peebles is in between), and the others lugged it up to the roof of Guyot Hall, the musty brick building on Princeton's campus that houses the university's geology department. But before they could get it running, they learned of another team of scientists who had evidently discovered the edge of the universe by accident. "A former graduate student of mine named Ken Turner heard a talk Jim Peebles gave in which he talked about what we were working on," said Dicke. "Turner mentioned it to Bernie Burke [a radio astronomer now at MIT but then at the Carnegie Institution of Washington]. And Burke told Penzias."

Penzias was Arno Penzias, a young radio astronomer at Bell Laboratories' Holmdel, New Jersey, research facility, no more than thirty miles east of where Dicke and his colleagues were working. Holmdel is not far from the Atlantic Ocean to the east and Raritan Bay to the north; from Crawford Hill, where a Bell Labs scientist named Art Crawford had built an antenna, the skyline of New York City is visible through the trees, just beyond the bay. The antenna is scoop-shaped and twenty feet across at its widest point. It was a prototype for what AT&T hoped would be a series of antennas for receiving and transmitting long-distance telephone signals. Wilson and Penzias were using this one to try and detect microwaves from the Milky Way. They were having a problem, though: there was a mysterious noise within the antenna—electromagnetic static coming from an unknown source. Noise is the bane of radio astronomers' (and communications companies') existence, distorting incoming signals and garbling any message, whether telephonic or cosmic, that passes through the system. In radio astronomy, the problem can be dealt with by understanding exactly what the source of the noise is, and taking it into account. But that can be extraordinarily difficult. The noise can come from almost anywhere. It can be radio transmitters in the area (Penzias and Wilson considered that their antenna might be picking up some spurious signals from the city nearby) or by atomic-scale vibrations within the electronic detectors attached to the antenna or even by debris lodged in the antenna itself (pigeons had roosted in the one at Holmdel, and Penzias and Wilson carefully cleaned out their droppings, to no avail). The astronomers wondered whether it might be coming from the Van Allen belts of charged particles that hover in the upper atmosphere, trapped by Earth's magnetic field. A high-altitude nuclear test in 1962 had filled the belts with extra particles, which might be causing the static. But the belts quieted down after a year or two, and the antenna noise persisted.

Penzias and Wilson had also considered that their noise might be coming from the universe itself, some form of cosmic radiation that appeared to come from all directions at once, which would explain the fact that it persisted no matter where they pointed their antenna. Their speculation might have ended right there except for Ken Turner and Bernie Burke. "When we heard about their noise," said Dicke, "we went over and had a look at their setup. We had kind of a problem convincing them that it was the same cosmic microwave background we'd been talking about. Their attitude was Let's get this noise figured out so we can go out and do some real astronomy."

The two groups finally decided that they would publish their work simultaneously in *Physical Review Letters*. Dicke's group described their theoretical predictions of what radiation left over from the superhot fireball of an early universe should look like, and Penzias and Wilson described the curious radiation they had found. That they matched was eventually understood to be powerful evidence that the universe had indeed been hot and dense at one time. This evidence, coupled with the fact that the entire universe is expanding, was enough to convince virtually all astronomers that the Big Bang did in fact happen. (The major competing idea at the time was that the universe had no beginning, and that it had always looked pretty much the same—the steady state model. The term "Big Bang," which was coined by the British astrophysicist Sir Fred Hoyle, who believed in the steady state, was originally meant to be a put-down.)

The evidence was convincing, but not entirely conclusive. The radiation still bouncing around the universe from the moment light could first break free should not come in just one electromagnetic frequency; like sunlight, it should be made up of many frequencies, mixed together in various strengths. It should have the characteristic signal of a "black body." A black body, in the jargon of physicists, is an object that is at perfect thermal equilibrium. It's called black not because it's necessarily dark—it can equally be brilliantly bright—but because it reflects no light; instead all the electromagnetic energy that falls on it is absorbed and diffused through it. The energy has to go somewhere, though, so the black body reradiates it, but whatever characteristics the radiation had when it arrived are transformed in the process of absorption and distribution. No matter what energy was used to heat it—radio waves, light waves, X-rays—the black body will reradiate the energy in a mix of frequencies that's solely determined by temperature. The composition of the body is irrelevant; any two black bodies at a given temperature will send out exactly the same mix of radiation.

After the Big Bang but before matter had thinned and cooled enough for radiation to move around, the early universe was one gigantic black body. The constant absorption and reemission of high-energy photons spread energy evenly throughout the universe, and when the light finally leapt free of matter, it bore the mark of that thermal equilibrium. The cosmic microwave background had the telltale characteristics of black-body radiation. Give physicists a temperature, and they could confidently say what the mix of wavelengths should look like. At the time of decoupling, the radiation was a black body shining at several thousand degrees with a unique mix of wavelengths. Now, billions of years later, it has cooled somewhat: it has become a black body at just about 2.7 degrees above absolute zero. (Absolute zero is the coldest temperature possible. What we call temperature is really a measure of the average energy of motion of the particles of matter. Even in a piece of rock the atoms, fixed in their places, are vibrating with energy at ordinary temperatures. Absolute zero is the point, $-273°C$ or $-459°F$, where even that motion stops.)

Since black bodies at different temperatures have their own characteristic mixes of radiation wavelengths, you can plot the intensity of radiation at each wavelength and get a curve, with a unique shape for every temperature. It's impossible to characterize a curve with only a single point, and so while Wilson and Penzias's detection of microwaves with a wavelength of seven centimeters might well fit on Dicke and company's predicted black-body curve, it also might not. You need two points on the curve even to begin understanding its overall shape. "So we went ahead with our radiometer," said Dicke, "and made another measurement at 3 centimeters. When Dicke did the matchup, the intensity of radiation at the second wavelength, too, was just what a 2.7-degree black body should be sending out.

Since then, astronomers have made thousands of measurements of the spectrum of the background radiation, at dozens of wavelengths. In virtually all cases, it has matched what physicists predicted a black body radiating at 2.7 degrees should look like. Even now, though, nearly thirty years later, they are still making measurements. It is not so much that anyone expects to learn that the universe was actually at a different temperature than we think at the time light freed itself from matter, or that the cosmic microwave background radiation isn't evidence after all that the universe was once very hot and very dense. In fact, the cosmic background explorer satellite, familiarly known as COBE and launched by NASA in 1989, has made a survey of the sky more sensitive and comprehensive than anything possible from the ground, and proven that the actual distribution of wavelengths is at worst .01 percent different from the ideal black-body curve.

When NASA scientists announced the result at the winter 1990 meeting of the American Astronomical Society in Washington, the audience broke into applause.

But there is still a chance that someone will find an anomalous bump somewhere on the curve, most likely at the long wavelengths that have not been explored very thoroughly. Such a discovery would not imply the failure of the Big Bang model; it would simply mean that sometime near the edge of the universe some other process took place besides the emancipation of light, some process that likewise was universe-wide and energetic. One possibility is the decay of some short-lived, ancient subatomic particles. Another is a generation of early, fast-burning stars. Another is a series of supernova explosions. There was in fact an experiment that found some evidence of just such an energy release.

While the space shuttle was grounded and COBE waited in storage for its chance to go into orbit, a team of physicists from the University of California, Berkeley, and the University of Nagoya, in Japan, sent a microwave detector aloft on a ten-minute suborbital rocket flight. Sounding rockets like this one and high-altitude balloons were, before COBE, the only way microwave astronomers could rise above at least part of the atmosphere and peer at the background radiation without the murk of water vapor getting in their way.

The Berkeley-Nagoya rocket found an apparent excess of radiation on the long-wavelength side of the curve that was a whopping 10 percent higher than theory said it should be. The discovery caused an enormous stir. Ten percent was a gigantic number and would have pointed to some terrific energy release, but subsequent rocket flights were unable to duplicate it and COBE later showed that there was nothing really there. The astronomers who made the original measurement decided that it was some sort of mistake—a glitch in the detector, perhaps.

If they had been right, the discovery might have been of Nobel Prize caliber. The original discovery of the background radiation itself certainly was. Penzias and Wilson got the prize in 1978 for their chance detection. Dicke and his colleagues did not get a thing. Some of Dicke's colleagues were offended. After all, the man who had realized there might be background radiation and set out to find it should get at least as much recognition as the two who found it by accident. There are physicists who, when polled every year by the Nobel committee for physics nominations, still submit Dicke's name.

There are two schools of thought about why the committee passed over the Princeton group. One is that, unintentionally or not, the Princetonians had really just reinvented an idea Gamow, Alpher, and Herman had had in the 1940s. Dicke was extremely clever but not original. In our conversation, Dicke acknowledged that this was a plausible explanation. "It also would have been difficult for them because there were so many of us involved. If they had given prizes here, Jim Peebles would certainly have gotten one. Dave Wilkinson, certainly. Possibly me as well, though I was really just the one who suggested what they should go out and look for."

Having made his fundamental contribution to yet another area of physics, Dicke moved away from the discovery and on to other crucial problems of cosmology. At the time we spoke, I asked him whether he was still working on the theory of an oscillating universe. The question was mostly polite; I presumed that Dicke's appetite for knotty problems had ebbed with his physical energy. "Not right now," he said, "but I intend to get back to it soon. I may have something to publish in a while."

David Wilkinson, on the other hand, never recovered from his encounter with the edge of the universe. His hair had grayed over the years, too, but compared with Dicke—compared with me as well, in fact—he glowed with good health. He was wearing a ski sweater when I first went to see him on a drizzly, cold December day, and in a year of further encounters I only saw him dressed up once, the night he introduced Margaret Geller to the audience at the public lecture at Princeton's Woodrow Wilson School. Wilkinson is an avid skier and motorcyclist. He and his wife sometimes go on long-distance rides out West. Where Dicke has a photo of

David Wilkinson   PHOTO: EILEEN HOHMUTH-LEMONICK

a lunar experiment, the most prominent thing on Wilkinson's wall is a ski poster, and where his old colleague's room is dominated by filing cabinets, in Wilkinson's office there is a huge drafting table covered with works in progress.

"That's where I design my instruments," he said in a soft, genial voice. "That's one difference in style between astronomers and physicists (and by training and temperament, I'm a physicist). Classically, astronomers didn't create their instruments. They just showed up and used them. Physicists have always designed and built their own apparatuses [a machine-shop course is required for Princeton's physics graduate students; the same is true at most other universities]. It's still rare, though no longer unheard of, in astronomy."

Since his 1965 work with Dicke's troupe, Wilkinson has stayed with the problem of measuring the microwave background, looking at both its spectrum and its isotropy—its evidently perfect smoothness. Currently, he's in a spectrum phase. "I'll tell you," he said, "that Berkeley-Nagoya rocket experiment had us worried for a while. The data were a real problem for us to explain, and we thought they were in trouble from the start. I was relieved when they turned out to be wrong. But I respect the people who did the work. These are really hard measurements. It doesn't take much rocket exhaust in the detector, for example, to produce that effect."

In the quarter century since the original measurements, Wilkinson has worked on dozens of microwave background experiments. "Do you want to see one?" he asked. "I have a graduate student who's testing out a new antenna we just built. She's going to do the actual observations down in West Virginia, but right now it's on the roof."

The roof of Jadwin Hall is flat and broad, the perfect place to do all sorts of physics experiments that require access to the sky, and the elevator goes all the way up. The view from the roof is panoramic; just uphill is the math tower, the tallest building on the Princeton campus, and a little farther up is Peyton Hall, the astrophysics building, with its observatory domes (the telescopes inside are used mostly for undergraduate stargazing). A little farther uphill and across Washington Road, I could see Guyot Hall, on whose roof Wilkinson, Dicke, and their group first tried looking for the edge of the universe.

The direct experimental descendant of that historic search was a few feet away from me, gleaming in the sun. It looked like an aluminum ice-cream cone about four feet high, its point held some four feet off the ground by aluminum scaffolding. There was a large box attached to one side, filled with electronic circuit boards; the whole thing dripped with cables. The scaffolding held the ice-cream cone so that it fitted at the bottom into a giant thermos bottle. The thermos (they're called dewars) had a hose attached, which led to a chugging vacuum pump.

Standing next to the apparatus and wearing a ski jacket and skater's headband against the February cold (for some reason,

Princeton's physicists tend be more robust and athletic than their counterparts in astrophysics, just up the hill in Peyton Hall) was a young woman with long blond hair. Her name was Suzanne Staggs, and she was the graduate student Wilkinson had sent me up to see. She had not yet been born when Wilkinson and the others went searching for the cosmic microwaves, and was surprised to learn that the building we were standing on had not been built at the time. The site had back then been an open field where ROTC cadets practiced marching, and alumni held cocktail parties on the tailgates of their station wagons before football games at Palmer Stadium, just to the east.

Staggs's thesis would come out of her attempt to use this antenna to measure the microwave background at especially long wavelengths, trying to fill in gaps at the far edge of the black-body curve. The reason for setting up the antenna in Princeton before taking it down to West Virginia was because, as Wilkinson had said, these measurements are very difficult. The chips and circuits

Suzanne Staggs   PHOTO: EILEEN HOHMUTH-LEMONICK

that make up the nervous system of the detector can fail in all sorts of ways, adding unwanted noise that confuses the issue; even the horn itself, if not machined with perfect accuracy, can throw the experiment off. Better to ferret out as many problems as possible here, where they can most easily be fixed, than out in the field. "We'll be testing all night," said Staggs, "and probably all weekend. We'll probably head for West Virginia in a few weeks."

As it turned out, they had to wait a few months. There was the half-expected, unexplained source of noise in the detector, and Staggs and the people working with her had to take it apart and fix it. By April, though, the major glitches had been ironed out, and Staggs and several helpers disassembled the antenna, packed it into a yellow Ryder rental truck, and drove down to Green Bank, West Virginia, where the National Radio Astronomy Observatory (NRAO) has several telescopes.

Green Bank is every bit as spectacular as Mount Hopkins, but in an entirely different way. You drive down a bucolic valley (I arrived around sunset, so the mist was beginning to rise), with the rolling, rounded Appalachian Mountains on both sides and farms in the bottomland in between. The road goes past herds of cows, and, in cow-less pastures, herds of deer. You pass a building with a sign out front advertising deerskin items: everything from clothes to gloves. Traffic is nonexistent, even though it is rush hour.

Suddenly, the road makes a sharp left hook and then a sharp right, and there on the right, in the middle of what looks like yet another pasture, is a dish antenna much like the satellite dishes that scoop up television signals and funnel them into houses along the valley. This one is probably thirty feet across. It is no longer used; in addition to being an observatory, the Green Bank site is an open-air museum of the history of radio astronomy. The thirty-foot dish, which sits next to the site's main entrance, was the home-built antenna made in the 1930s by Grote Reber (he was nicknamed "the Wildcat Astronomer"). Reber was actually an electrical engineer who did radio astronomy as a hobby. He set up that first dish in his hometown of Wheaton, Illinois, a suburb of Chicago. His neighbors thought he was using it to make rain. Reber was the first one who proved radio waves could be useful in studying the universe, just as light waves had been.

On the other side of the entrance road was a white-painted metal frame, shaped somewhat like the trailers that haul new cars down the highway, It was a full-size re-creation of the radio antenna Karl Jansky of Bell Labs was using when he detected the first radio signals from space in 1931—a wholly unexpected and unpredicted phenomenon. Like his Bell Labs colleagues thirty-four years later, Jansky was testing an antenna the company was considering using for telephone transmissions, and he kept getting static in the receiver. In his case, the static was periodic, not constant; it appeared at about the same time each day, then went away. After observing the static for several years, he finally realized that he was receiving radio waves from the Milky Way itself.

The working radio telescopes on the site dwarf these two relics. Beyond the NRAO administration building, which is just inside the entrance, is an 85-foot radio dish, and beyond that is a 140-foot monster. The puny antenna Wilkinson and Staggs had dragged down here was too small to be visible; all I could see was a tiny splash of white—the truck trailer they were using as a control room. The reason they had come all the way to Green Bank was the same reason the NRAO has its antennas here: the valley is a radio quiet zone. No one in the area is allowed to broadcast radio signals that might interfere with the telescopes' operations (the military is an exception, but even they inform the site director, after the fact, of what frequencies they have been using in various exercises). Just as the astronomers near Tucson have convinced city authorities to install streetlights that are minimally troublesome to the telescopes on Kitt Peak and Mount Hopkins, the NRAO has negotiated with the Federal Communications Commission for the authority to tell people to get off the air.

The Princeton group had been at Green Bank for two days when I arrived, and as I drove in, dinner in the cafeteria was just ending.

Staggs was sitting at a table, which held four fifths of her team: Ed Wollack, another graduate student; Norm Jarosik, an instructor in the physics department; and Ted Griffith, who runs the shops in Jadwin Hall. A nonphysicist, Griffith has been on more trips to exotic observatory sites than most Ph.D.'s. As Wilkinson told me, physicists not only make many of their own instruments but they also are often responsible for the structures that hold them. It pays to have someone along who knows construction. Griffith has accompanied physicists to Australia, to the National Scientific Balloon Facility in Palestine, Texas, and to the South Pole. Griffith and Jarosik had arrived independently, the former on his motorcycle and the latter in his airplane, landing a quarter mile away at NRAO's private airstrip.

Only Wilkinson had not arrived yet—he was at a meeting at the Goddard Space Flight Center in Greenbelt, Maryland, COBE's home institution, talking about the latest results from the satellite and about future space-based microwave measurements. He was

Ed Wollack   PHOTO: EILEEN HOHMUTH-LEMONICK

scheduled to arrive later that night, but the antenna crew couldn't wait to show off their handiwork. We piled into a beat-up light blue Checker diesel station wagon, the property of NRAO, and headed down the road toward the towering 140-foot antenna. Halfway there the road was blocked by a pair of railroad-crossing-type barriers topped with red lights; just beyond them was a sign that displayed a red circle with a red diagonal line going through it and a picture of a spark plug in the middle: the international no-spark plug symbol. At optical telescopes, it is stray light that can ruin an observation, but here the danger is stray radio waves. The tiny electrical discharges in a spark plug broadcast static, like miniature lightning strikes. Beyond this point in the road, only diesels, which have no spark plugs, and bicycles—of which the observatory has a score, available for anyone to use—are permitted.

The mist was getting thick, and Norm, pointing to it, said, "Hey, who left the nitrogen open?" When it is operating, the microwave detector must be cooled in a bath of liquid helium to four degrees above absolute zero ($-269°C$ or about $-452°F$). Otherwise, vanishingly faint electromagnetic signals from heat vibrations in the detector's molecular structure would overwhelm the even fainter whisper of radiation streaming steadily in from the universe. The liquid helium in turn sits inside the dewars, and rather than pouring expensive liquid helium into the room-temperature dewars—which would freeze the air trapped inside and contaminate the container—the physicists pump cheap liquid nitrogen, at about seventy degrees above absolute zero, in first. The nitrogen is later flushed with helium gas and, finally, the liquid helium itself is pumped in from an even bigger dewar sitting on the platform. The apparatus was now in the liquid-nitrogen stage. Liquid nitrogen cools the air and creates a fog when it's spilled; Jarosik was making a physics joke.

About a quarter mile before it reached the 140-foot dish, the Checker turned right and headed a short way down a gravel road to the experiment. The aluminum horn looked just as it had on the

Norm Jarosik   PHOTO: EILEEN HOHMUTH-LEMONICK

roof of Jadwin, like a rocket poised to take off straight down to the center of the Earth. It sat on a low wooden platform, built for the occasion by local carpenters, and surrounding the platform were six four-by-six-inch beams, fourteen feet long, set into the ground like telephone poles and arranged in a hexagon. These would hold the ground screen, a mesh of ordinary chicken wire, doubled and strung over wires suspended between the beams and the platform. The ground itself radiates microwaves that would contaminate the signal from space if the detector even glimpsed them sidelong, and so Griffith would supervise (and do much of the work on) the building of a bowl-shaped net of wire—the ground screen.

Perhaps fifteen feet from the edge of the platform was the white trailer. Inside were two computers, two desks, and drawer after drawer of tools, solder, bolts, and such. When Staggs and the others had arrived a few days earlier, the trailer, on loan from NRAO, was filthy. There was unmistakable evidence that rats had been camping out in it. Now, after a run to the local grocery store for emergency supplies, the place was tidy and reeked of Pine Sol. The equipment seemed to be in good order, so the crew headed back to the dormitory, where the evening's entertainment would be a film festival of sorts: three documentaries prepared by NRAO about its work.

The movies were of the genre that used to be shown (and for that matter, probably still are) to junior high school science classes. They were light on what people in Hollywood call "production values," and heavy on pompous narration. The audience alternated between howls of laughter at the style and frank amazement at its content. One film was about Grote Reber. Another documented the construction of the 140-foot radio dish in the 1950s. At one point while the narrator talked about the delicacy the construction crews had to use in assembling the multiton sections of the 140-foot radio telescope, the picture on the screen was of a worker banging a section into place with a large sledgehammer. It was unclear whether this was meant to be funny. Staggs, Jarosik, Griffith, and Wollack are people who build things, though, and they could appreciate the engineering and construction expertise that had gone into NRAO's inventory of mammoth instruments even through the haze of bad moviemaking.

Wilkinson didn't arrive until late, and the full crew didn't assemble until breakfast the next morning. Wilkinson was anxious to get caught up. "I was going to bring a telescope so we could align the antenna with the North Star, but I forgot. Well, I have a gunsight in my pickup that we can use. Is the platform okay?"

"It isn't level," said Griffith. "It kind of bows in the middle."

"Well, is it solid?" asked Wilkinson.

"Oh, yeah."

"How's the trailer?"

Suzanne told him the story of the rats. "They even chewed through the electrical cables, so it had to be rewired."

"Next time," said Jarosik, "we'll leave the power on. They'll be sorry."

Breakfast over, the scientists headed down to the observing site for a day's work. When engineers have finished the design stage of a project and manufacturing is about to begin, they say they're ready to "bend metal." These physicists were ready to bend metal without the quotation marks. The microwave antenna was all set up, but there was still the ground screen to build. It was back-breaking work, and in between checking out the computers and other equipment, all of them, Dave Wilkinson included, helped haul rolls of chicken wire around, and cut them to size, cutting their hands in the process.

Although she was junior to both Wilkinson and Jarosik in her academic status, this experiment was Staggs's, and for now they were working for her without any apparent ego problem. "What can I do to help you out?" asked Wilkinson. His first job was to mount his gunsight on the antenna's scaffolding. He found a small, flat piece of metal, bent it so it formed an L shape, and drilled several holes in it. This would be the mount. "We need to know where we're pointing," he explained, "because there's a certain part of the sky we want to aim at." If the horn pointed straight up all the time, he explained, the thin band of the Milky Way would eventually pass overhead. As Jansky discovered, the Milky Way fairly crackles with radio noise, which would contaminate the signal from deep space. So the physicists would tilt their horn antenna to point at the "cold spot"—ninety degrees away from the Milky Way and thus through the least possible interference. As the cold spot moved into a different part of the sky with the Earth's rotation, the antenna would be retilted. To make it as easy as possible to follow the cold spot, the entire structure would be rotated to point due north and then tilted (the six legs of the scaffolding had wheels at the bottom).

Even in the cold spot, the sparse molecules of interstellar space add a half a degree's worth of microwave temperature to what the detector sees. The aluminum antenna itself adds another half degree. And the atmosphere adds two degrees. The last makes it doubly sensible to tilt the apparatus. When it aims straight up, the horn is looking through the thinnest possible layer of atmosphere. When it's tilted toward the horizon, it is looking obliquely and thus is passing through a thicker chunk (that's why the setting sun is often red but the noontime sun isn't). That helps them double-check the atmosphere's microwave temperature. The horn looks at an enormous section of sky at a time, eighteen degrees across, or thirty-six times the width of the full Moon, so the aiming need not be terribly precise. "We do have to know which way north is, though not too accurately," said Wilkinson. Then, a perfectionist like most experimental physicists, he did it accurately anyway, spending a good half hour aligning the gunsight so that it and the telescope were aimed at exactly the same point. Later that night he and Jarosik would spend a comparable amount of time squinting through the gunsight and nudging the detector so that the North Star was lined up precisely in the sight's crosshairs.

Meanwhile, Griffith was pondering the problem of the ground screen. This was hardly the first time microwave astronomers have had to shield their experiments from stray emissions, but he seemed to be designing this one from scratch as he went along. That afternoon, sitting on the ground and sewing pieces of the screen together with stainless steel wire and a pair of needlenose pliers, Wilkinson explained why. "We're looking at much longer wavelengths than we ever have. Ted built us a screen up in Saskatoon, Saskatchewan, when we were observing one-centimeter radiation, but here we're looking at around twenty-one centimeters. That takes a completely different kind of screen."

Twenty-one centimeters is a magic wavelength in astronomy. It is the broadcast frequency of single atoms of hydrogen; when they are disturbed by incoming radiation, the electrons circling hydrogen nuclei flip in place like tiny compass needles suddenly jumping from north to south. When they flip back again, a fraction of a second later, they send out a tiny ping of radar, of microwaves. The accumulated radar beacons of uncountable atoms betray the locations and movements of giant hydrogen clouds in the Milky Way and other galaxies. By international agreement, this particular frequency and those just around it are off-limits to terrestrial broadcasters. Like Green Bank, twenty-one centimeters is a radio quiet zone. "Actually," said Wilkinson, "we skip over twenty-one centimeters itself because the Milky Way is so bright at that wavelength, even in the cold spot, that it would overwhelm us. But we look at all the wavelengths around it."

Later that evening, Wilkinson and Jarosik went outside to adjust the electronics—the circuits that amplify, process, and record what the detector sees—and attach a portable oscilloscope to monitor what was going on. The oscilloscope, translating the electronic signals from the detector into lines on its built-in television screen, showed the basic signal from space—a jagged, horizontal line. Occasionally, every eleven or twelve seconds, a much more irregular line would flash faintly on the screen for a half second. "That's radar," said Jarosik, "from an airport somewhere." Such interference is normal and easy to deal with. But Wilkinson, Jarosik, and Staggs, inside at the computer, were on the trail of something more confusing. "This was bothering us back in Princeton," said Wilkinson, "but then it went away. Now it's back." The problem was a half degree's worth of noise in the antenna that could not be accounted for. "It's not much, but you have to understand what it is. That horn is a complicated thing. It has wonderful properties, but it has a complicated shape inside."

The shape—ribbed and tapered—is designed to funnel microwaves down into the detector. But it could be that waves of a particular frequency were getting trapped in the horn, rattling back and forth without being able to escape, and heating the metal just slightly. Would that require remachining the whole thing? I asked. "No," said Dave. "I'd just put a C-clamp on it. I'm not proud." The clamp, he explained, would interrupt the bouncing wave.

As it turned out, the clamp was not needed; Staggs and Jarosik stayed up until five o'clock the next morning, and were back at the detector before lunch, but they finally traced the half degree to a faulty seal between two sections of the horn. A tiny bit of heat was seeping in from the sides. It took a hastily fashioned gasket—an O–ring—formed of solder and squeezed between the two sections to stop the leak. While they were puzzling out the problem, Wilkinson talked about what the experiment was going to try and accomplish.

The COBE satellite has already established that at its most luminous wavelengths, a few millimeters long, the microwave intensity does indeed match the theoretical curve of the background radiation to within a fraction of a percent. "What it means is that the Big Bang is an awfully good theory," said Wilkinson. "There is no known process other than the Big Bang that could create such a curve, and it is clear that not much has happened since then to distort these Big Bang photons."

But "not much" is different from "nothing at all," and in fact there are a number of things that might have happened to put a little bump at one location or another along the otherwise perfect black-body curve. Nothing so gross as what the Berkeley-Nagoya experimenters thought they had found, but something. When the first stars formed, for example, their radiation could have ionized much of the universe's gas, stripping electrons from atoms, and that gas would then have regained the power to knock photons around. This would show up in the detector as an unexplained dip in the black-body curve at long wavelengths; the photons that ordinarily would have come straight through to Earth were scattered another direction instead, and there would be a deficit in photons of a particular wavelength. "We're talking on the order of one percent," said Wilkinson. "This experiment covers the longest wavelengths this group has ever studied," he continued.

"Other groups have looked at long wavelengths and haven't found anything, but their error bars were pretty big."

The other question, the one that bears more directly on the crisis in cosmology, is that of anisotropy. No one had yet seen concrete evidence of temperature fluctuations, even minor ones, in the microwave background from one point on the sky to another. Yet the seeds of the modern universe's structure—the small fluctuations in density that given ten or twenty billion years of gravity grew into Great Walls and Great Attractors—should be visible as just such a pattern.

"The point is," said Wilkinson, "that at the epoch of decoupling, matter and radiation were interacting strongly. A photon I see today coming from a particular part of the sky must have had a last scattering, a last encounter with an electron. In the simplest universe, it came right from that last interaction straight to me. Now, suppose the region it came from was more dense than another. That is just another way of saying there was more matter there. And that means there were more electrons to scatter photons one last time before they decoupled, and thus more photons scattered. Here's another way to look at it: all other things being equal, a higher-density region is by definition hotter than a lower-density region. The temperature of the anisotropies tells you the density of the clumps, while their angular size on the sky tells you the size of the clumps.

"We're pretty sure we've got to see these temperature fluctuations, and that's where the crisis is coming from. We see clumps of matter today, so we should see clumps of radiation, and based on the size of clusters today those clumps of radiation should be on the order of ten arcminutes across." An arcminute is one sixtieth of a degree. The half-degree-wide full Moon is thirty arcminutes across.

There are other effects that should make the cosmic microwave background anisotropic as well. All of them are related to the distribution of matter in the universe both before and after the era of decoupling and recombination. One phenomenon is known as the Sunyaev-Zel'dovich effect, named for the Soviet astrophysicists who first described it (Zel'dovich was also the one who claimed the perturbations in the early universe had to be the same on all scales). The earliest concentrations of matter, the incipient galaxies or clusters of galaxies, were probably pervaded with hot, ionized gas, energized by the collapse of proto-galaxies under gravity, and the gas would have bumped some cosmic photons to higher and thus unobserved frequencies; the microwave background, evidently missing some photons, would look cooler at those points. Or the temperatures could be elevated instead, as electrons in those primordial clusters collided with each other, generating what is known as bremsstrahlung radiation to boost the energy of the background. Or there might have been turbulence. Basically, anything that added energy unevenly could have done it.

There should be deviations as well at much larger scales than the ten arcminutes associated with primordial clusters caused by a phenomenon called the Sachs-Wolfe effect. Photons of the cosmic microwave background, moving through areas of the universe with a higher than average density of matter, will be fighting a higher gravitational field than average as well. And gravity can force electromagnetic radiation to redshift even if the source of radiation is not moving away from Earth. On scales of a degree, explained Wilkinson, you'd statistically expect some regions to contain slightly more areas of high density than others, even if the overall average across the sky is uniform. Those one-degree regions should be slightly warmer than others. And now the theorists have calculated that if Great Attractors exist in any significant numbers, they themselves should directly produce one-degree variations on the sky. Finally, the energy added to or subtracted from the microwave photons in collisions with electrons can produce red- and blueshifting too.

The anisotropies resulting from all these processes end up being superimposed on one another, producing a pattern that is extraordinarily hard to decipher. "It's in a sense irrelevant which scale you use if you're looking for anisotropy," Wilkinson said. "The theories that predict it at large scales are just as fundamental as those that predict anisotropy at small scales." Inflation, in particular, says that fluctuations in the early universe should have about the same average amplitude at all scales—the Harrison-Zel'dovich-Peebles power spectrum. The consequences for fluctuations in the microwave background should be that at large scales, the anisotropies, the deviations from the average temperature, should be about the same when you look at two different patches of sky—any two patches at all, as long as they're the same size. The theorists who analyze the signals from all-sky microwave detectors like COBE look for a dipole anisotropy, which is dividing the sky in two halves and comparing their temperatures. They look for a quadrupole, which is cutting the sky in quarters and comparing these. Then they look for an octupole—eight equal parts—and so on, until the fractions of sky they're comparing are too small to be contained within the antenna's beam.

This explains why the COBE satellite is valuable even though it has a seven-degree-wide beam and thus can't probe anywhere close to the scale of the imprint of individual structures—even superclusters of galaxies—on the cosmic microwave background. "For one thing COBE scans the whole sky," said Wilkinson, "and not just a few spots. And it does it over and over, so you can be a lot more confident about the measurements. Second, although that scale is too big to spot even the seeds of the Great Attractors, the Sachs-Wolfe effect can be measured on that large a scale. It's fair to say that at all angular size scales, there is really good work going on. The real problem with COBE is that there's an enormous amount of instrument noise. It will be hard to get the noise down to the point where we can detect anisotropies in the interesting range of a few parts in ten to the sixth. There's so much to do in this field that it's hard to know what to do next.

"Of course," said Wilkinson, "when you're designing an experiment you always have to make a choice of what angular scale to use. The ultimate dream is a map of a ten-degree-by-ten-degree sector of sky, showing all temperature variations down to ten arcminutes in angular scale and with temperature sensitivity of one part in ten to the sixth. *That* would give us a complete power spectrum and really put inflation to a direct test."

I suggested that these concepts were difficult for most people to grasp. "Listen," said Wilkinson, "I've been thinking about these things for years, and I still don't understand all of it. In 1965, no one was looking for anisotropy—we were much more interested in finding large-scale isotropy, smoothness, to show that the microwave background was coming from all over the universe, not just the galaxy. When we went up on the roof, we were measuring how round the universe is. Early on, Peebles did some calculations of what temperature variations should be seen in the microwave background based on the density variations in the modern universe. Back then, though, the question was Big Bang versus steady state . . . there was some vague, muddy thought that you were probing the early universe, but the real issue was just establishing that the microwave background really was what we thought it was. Juan Uson and I did look, in about 1980, for anisotropy at one-centimeter radiation at a resolution of five arcminutes. We didn't find any, and that was already shaking up the theorists."

The next day was Sunday, the day the cafeteria is open to anyone in the local community who cares to come, and, thanks to the government-subsidized meals, plenty did. In contrast to the other meals when the five members of Staggs's crew and two Japanese astronomers working on the 140-foot were the only diners present, the room was filled with locals, still dressed up from church and ready to try the fried chicken special. I asked the crew from Princeton whether the astronomers ever mingle with them. "No," said Griffith. "They're afraid of us, and we're afraid of them."

I left Green Bank that afternoon and headed back to Princeton. The experiment would be running for the next year or so, and the physicists assured me that no results would be available for at least that long. As it turned out, it would be longer than that. Staggs told me later that many of the measurements they went on to make that spring were contaminated by moisture—dew that had condensed on the plastic wrap shielding the opening of the antenna. In addition, the eighteen-degree-wide beam was so big that microwave emission from the atmosphere was hard to distinguish from the cosmic signal. Staggs and Wilkinson would eventually remove their detector from the apparatus altogether and bring it back to New Jersey, where they would attach it to an antenna with only a two-degree opening: the original horn that Wilson and Penzias had been using when they first saw the edge of the universe. Bob Wilson, still working as a radio astronomer (while Penzias had risen to become a Bell Labs vice president) would be their collaborator.

None of that was predictable as I headed away from Green Bank to keep an appointment to see Lyman Page, another Princetonian, also a physicist and also on the trail of the cosmic microwave background. Sometimes the microwaves are measured in the bucolic peace of Green Bank; sometimes in the suburban comfort of the Goddard Space Flight Center, where the COBE control room is located; and sometimes in less benign conditions. Page has spent much of his career working in the last category.

Although his office is in Jadwin Hall, just down the corridor and around the corner from Dicke, Peebles, and Wilkinson, and although other physicists had been suggesting for weeks that I talk with him, what finally triggered our conversation was a chance meeting on the front sidewalk of the Carnegie Family Center outside Princeton—the day care center where my daughter used to expend much of her phenomenal energy. Most mornings as I dropped her off I would see a tall, blond-haired young man with glasses and a case of academic rumple walking a bicycle with one arm and carrying a baby boy in the other. When we finally introduced ourselves, after several months of smiles, I recognized his

William *(top)* and Lyman Page   PHOTO: EILEEN HOHMUTH-LEMONICK

name at once and began talking to him on the spot—with his son gurgling and my daughter grabbing me around the knees—about the edge of the universe.

It was clearly not the best time or place, which was why we made the appointment to reconvene in his office. His work load was crushing, but he was having a blast. "Basically, I come here and work all day, go home for dinner, and come back," he told me. "It's really tremendously exciting. The smoothness of the microwave background is a real problem, on all angular scales, and we're now close to ruling out all plausible theories about how the universe evolved."

Back in undergraduate school at Bowdoin College in Maine, Page had been particularly interested in Bob Dicke's career; he followed it with a kind of awe. "I read everything by and about him that I could," he said. "I knew someone who had been a postdoc under Dicke, and he told me they'd be sitting there talking, and Dicke would suddenly have an idea, and my friend would think 'Wow, I could *never* have an idea that profound.' And this would happen frequently."

Page was now carrying on the work that Dicke helped begin; in contrast to David Wilkinson (in his current phase) and Suzanne Staggs, he was working hard on trying to find out just how smooth the microwave background is. In his opinion, it was already smooth enough to mean the standard model was in real trouble. I asked him whether it was depressing to see a model of the cosmos delicately cobbled together over the past several decades come apart. "Depressing?" he asked, looking genuinely baffled. "It's really exciting, and I'm lucky to be here at Princeton right now. There are really just a few places in the world where people do this kind of work; it used to be that anyone could set up a radiometer and take good data, but the techniques and equipment have become so refined that just Princeton, Berkeley, Santa Barbara, and two places in Europe and one in the Soviet Union are doing any important observing."

For the long-wavelength microwaves that Staggs and Wilkinson were looking at, Green Bank was a perfectly good location, but for the short-wavelength radiation where the cosmic background shines brightest, there is one spot on Earth that is better than any other: the South Pole. "When you're observing at short wavelengths, water vapor is a big problem," said Page. Water vapor is like a sponge for short-wavelength microwaves. It absorbs them and converts their energy into heat (that's how a microwave oven works—the water in food absorbs microwaves and heats up, essentially steaming the food from within).

The Earth's atmosphere is full of water vapor, but the South Pole is less plagued with it than most places. One reason is that it's at very high altitude. The ice cap that covers the Antarctic land mass is more than ten thousand feet thick at its center, and the altitude of the Amundsen-Scott South Pole station, the permanently inhabited U.S. base is over nine thousand feet high (the ice is so heavy that it has pushed the rock underneath well below sea level). Antarctic veterans speak of going "up to the Pole" when they travel from McMurdo station, near sea level on the shore of Ross Island in the Ross Sea. The altitude puts polar scientists well above most of the water vapor in the atmosphere, and the intense cold forces most of what is left to freeze out into tiny ice crystals.

Lyman Page had been up to the Pole six months before our conversation, and he was preparing to go back for another season. "I got there the first week in November," he said, "and left the day before New Year. It was very tough. My son was only weeks old and it was hard to leave Princeton. But my wife and I had known long before he was born that this was something I'd have to do, so we were prepared."

The Pole is the closest thing on earth to space, as far as microwaves are concerned, and while unmanned balloon flights—which Page has also been involved with—are even better, reaching tens of thousands of feet, they are short-lived. A microwave detector at the Pole can operate continuously for weeks and months, which is ideal for measuring whatever deviations from perfect smoothness might be present in the cosmic background. For one thing, the detector points again and again at the same spots of sky as the Earth turns under it. The repeated measurements let the device check and recheck its own measurements. The chance that a jolt of static could be mistaken for a whisper from the edge of the universe is minimal; the jolt would only show up on one pass and no other. Another advantage of the Pole is that it is surrounded for hundreds of miles on all sides by a flat plateau of ice. There are no mountains, no valleys to roil the flow of air. Storm systems, when they pass through, do so quickly and smoothly.

The high altitude is a problem for scientists, though. When they first arrive at the Pole it takes three or four days, on average, to get over the light-headedness, shortness of breath, and rapid fatigue brought on by thin air. I had been to the Pole once myself a year before Page went, and I experienced the problem firsthand. I had been traveling with a group of reporters being shown around Antarctica by the National Science Foundation (NSF), which is in charge of all U.S. science on the frozen continent (science is in fact the primary activity for Americans in Antarctica). Because the Sun is up twenty-four hours a day during the Antarctic summer, planes fly around the clock. Ours got in from McMurdo at three in the morning, and the visit would end abruptly at three the next morning, when, after two or three hours of sleep we were awakened and told that there was a flight coming in with new personnel and they needed the beds. We would be departing in twenty minutes.

As we got off the plane, the Sun, shining off the snow pack, was blinding. We were allowed four hours of sleep and then, after a hurried breakfast, began a tour. The guide was John Lynch, the NSF's senior representative at the station, and he was clearly proud of his baby. We never stopped moving for the next ten hours except to have a fifteen-minute lunch and to have scientists explain their experiments. There was a group doing cosmic microwave background measurements that year, too, with representatives from Italy and from Berkeley. We were so exhausted and disoriented by the altitude and cold (due to a heat wave, it was ten below zero Fahrenheit) that nobody could follow a thing that they were saying, except for Malcolm Browne, the *New York Times* reporter who was utterly fascinated by every gadget he saw and conscientiously took notes on whatever the scientists told him, with cotton-gloved fingers that must have been numb.

John Lynch was so concerned that we see every last experiment and meet every last scientist that he was a little annoyed when I pressed him to let us take a break and walk over to the actual South Pole itself. "Okay, okay," he finally said in midafternoon. "Let's go over and you can take your hero pictures for the folks back home." A few hundred feet away from the dome covering the main buildings of Amundsen-Scott station is what looks like a barber pole stuck into the snow, with a gleaming metal globe on top, surrounded by a circle of national flags—the flags of countries who have signed the Antarctic treaty that governs all activity on the continent.

Most visitors assume this marks the South Pole, but it does not. The real Pole, the geographic Pole, was, at the time of our visit, about thirty feet away, marked by a simple surveyor's stake and by a sign. The barber pole marks the *ceremonial* South Pole. The reason there are two poles is that the entire ice cap moves, flowing slowly under its own tremendous weight. The geographic Pole, marking the axis about which the Earth spins, is stationary, and everything on the surface moves past it at about thirty feet per year. When it was built in the 1960s, the Amundsen-Scott station was deliberately built a quarter mile from the geographic Pole, so that the ice would slowly bring the station closer and closer to it. The idea was that the pole would lie right inside the dome in 1991 when the Antarctic treaty nations met there for their thirtieth annual conference, but the ice didn't move as fast as predicted. Because the original distance was too far to walk for some of the visiting VIPs—senators and the like—who periodically visit the station, the NSF built a ceremonial Pole for hero pictures. The ceremonial Pole has approached its physical cousin over the years (the stakes from each year's survey have been left in place, and you can see them marching off into the distance), and they are so close together now that the whole thing seems a little ridiculous.

The buildings underneath the station dome are essentially the same as refrigerated railroad cars—big rectangular boxes with thick walls and bulky doors—except that their job is to keep cold out, not in. The dome itself is there just to keep the snow off the buildings, and it is unheated, but station scientists and support workers usually don't bother putting on coats just to go from one building to another, even when the temperature drops to forty below, as is typical in the summer. The buildings house the kitchen and dining room; a library; a weight room; a small store where they sell liquor, stamps, and South Pole T-shirts; a medical dispensary; and dormitory rooms. The rooms are tiny and cramped, and visiting scientists—Page among them—are often happier to be assigned to Jamesways, insulated canvas Quonset huts that sit just outside the dome.

"The routine when I was down there was pretty simple," he told me. "You got up and walked over to the dome for a cup of coffee, then you commuted out to the experiment, which was a few miles from the base, and worked for ten or twenty hours." Much of the work is outside, setting up microwave horns, building ground screens, filling Dewars with liquid helium. There is also an on-site Jamesway, filled with test equipment, cots, and a small stock of food. "When you're really working hard, you don't want to take the time to drive back in just to eat and sleep," said Page. "Once in a while, you go in and see a movie on videotape, but other than that and meals, I stay away from the dome pretty much."

It wasn't Page's first visit to Antarctica. Back in the early 1980s, he went to McMurdo and stayed seventeen months, tending a cosmic ray experiment for someone else. "I had just graduated from college with a degree in physics," said Page, "and I didn't really know what I wanted to do. I heard about this job and I got it."

Cosmic rays are electrically charged fragments of atoms that fly through the galaxy, some of them smashing into the Earth's magnetic field. Magnetic fields tend to deflect charged particles, so relatively few of them reach the surface at low latitudes where the magnetic field arches high above the planet's surface. But in the Arctic and the Antarctic, the field dips to the surface. A compass needle right at the magnetic South Pole wants to align itself with the field and point straight up into the sky. And a cosmic-ray particle has a good chance of reaching a detector at, for example, McMurdo station.

"I went down in October, and stayed seventeen months," Page said. "There wasn't much actual work to do on the detector. I got it down to about ten minutes a day, and my hut was isolated from the rest of McMurdo station." Nowadays, the winter population of the station is about four hundred scientists and support people (the weather service, air controllers, and pilots are all supplied by the navy), and it rises to nearly two thousand in the summer. Back then, said Page, there were only fifty who wintered over at McMurdo, but Page had little contact even with them. "For a while," he said, "they had a small nuclear reactor up on Observation Hill just above the town to provide power, and the neutron radiation would have screwed up the detector completely. So we had the hut built about a mile from town, about halfway along the road to Scott Base [New Zealand's principal scientific station], where the bulk of the hill itself shielded us from the reactor."

During the summer, it would have been an easy walk into town along the gravel road, but in the nonstop darkness of winter, with temperatures more than fifty below and storms blowing up with little notice, Page only came in every other day or so. He didn't mind at all. "It was just spectacular," he said. "There were auroras—southern lights—probably seventy-five percent of the time. You could tell when it was a really good one, because they had a dogsled team at Scott Base, and the dogs would start howling. It looked like colored icicles coming down from the sky and touching the snow, shimmering and shifting around.

"When it was clear and the Moon was up, you could see the volcano Mount Erebus towering over the north end of Ross Island, with its plume of gases rising into the air, and the peaks of the Transantarctic Mountains across McMurdo Sound. The sky was crystal clear when it wasn't storming, and the stars were unbelievable. The red stars were really red, and the blue stars were really blue. One of the more amazing sights was just before the Sun came back up in the spring. It was still below the horizon, but the plume coming from Erebus was high enough that it was illuminated. Another one was when the ice on the bay broke up. One day, a half mile of ice just disappeared.

"The storms were great, too—just incredible. The hut was made of two layers of three-inch plywood with insulation sandwiched in between, and the whole thing was anchored to the rock with cables. There was no way the wind was going to blow it away, but it stretched the hut so that adjoining panels would separate, and the snow would blow in. I only went out in a storm once, when the smokestack blew off my heater. I repaired it in steps—I'd get ready to drive a nail, then run outside and drive it, then run in and prepare for the next step."

"The whole level of social interaction that you need around other people and which takes so much energy was gone. I did spend a lot of time with one guy, playing chess and cards, but I also spent a lot of time reading and thinking. I read *Moby Dick,* the Feynman lectures on physics, *War and Peace.* I also painted an abstract scene on the ceiling—a sunset, with cosmic ray showers coming down to hit the Earth. It was an experience I wouldn't have missed, but once was enough. I still enjoy going down for a couple of months, but I wouldn't go for a year."

When he got back from Antarctica, Page wasn't quite ready to reenter society. He hadn't yet decided what he wanted to do. He needed some time, believe it or not, to think. "I had saved plenty of money in Antarctica, since there wasn't much to spend it on, so I bought an old thirty-six-foot wooden sailboat in Maine. You could see through every plank, but I fixed it up and sailed down the East Coast and into the Caribbean. Friends would come along for periods of time to crew for me. We'd sail for a while and then find some work for a while, then sail again. We were headed for the Panama Canal, intending to sail for the South Pacific, when we got caught in a terrible storm and lost our rudder, a sail, and a bunch of rigging. I decided that night that instead of going to the South Pacific I'd go to graduate school in physics.

"We limped into Jamaica, got the boat fixed, and headed back north by way of Belize, Guatemala, and the Yucatán. When we got to Boston I got off the boat and went in to MIT to talk to Ray Weiss in the physics department. He gave me a job testing transistors, and I studied on the side for the graduate record exam. I did okay, and Ray persuaded the department to let me in. He has a reputation for taking on the occasional oddball case, and I was it that year. I ended up giving the boat to a lobsterman in Maine who had helped me fix it up. I couldn't possibly have done it without him and he really loved it, so I figured it should go to him."

After Antarctica and the Caribbean, I asked him, weren't Cambridge and Princeton a bit tame? "Well, I do miss the mountains and the sea," he said, "and I was a little bit worried that I'd hate it here. But I'm so involved with my work that it doesn't matter."

It's no wonder Page is so busy. Although he had recently returned from Antarctica when I spoke to him, he was busy analyzing data taken from an entirely different microwave experiment he'd taken part in a year and a half before, which involved a series of balloon-borne radiometers launched from the National Scientific Balloon Facility in Palestine, Texas.

"Actually, because of the prevailing winds up at 140,000 feet or so, where we do the launches depends on the time of year. If you're launching in spring, you launch from Palestine, and let the balloon blow west. If you're launching in the fall, you launch from Fort Sumner, New Mexico, and let it blow east. In the fall of 1990, we sent up an experiment from New Mexico in the evening—you always do it at night because you don't want the Sun shining into the experiment and frying it. We woke up the next morning and saw the balloon glinting in the rising Sun. At that point, it was over Oklahoma. It was an incredible feeling."

Aboard the balloon was a microwave horn, very similar to (except smaller than) the Green Bank detector Suzanne Staggs had set up, which dangled from the balloon and pointed at a forty-five degree angle from the vertical. A small motor spun it slowly like a record turntable so that every forty seconds it measured the microwaves coming in from a circular patch of sky. The turning of the Earth and its atmosphere with respect to the heavens means the radiometer recorded a different patch on every circuit, and overnight the horn mapped a full half of the Northern Hemisphere sky.

That's why the experiment is done in both fall and spring: every six months, the Sun appears to be in the opposite half of the sky (it's really that the Earth is on the opposite side of the Sun, of course). Thus, two flights a year will cover the whole Northern Hemisphere sky. "Our first flight lasted only eight minutes," said Page. "The balloon burst." But that was followed by two successful all-nighters, and the radiometer sent back so much data that Page and his collaborators, Ed Cheng and Steve Meyer, have been overwhelmed ever since. "It will take us a couple more years to reduce the data," he sighed. "Right now, I'm the only one working on it. I'm trying to get a grad student to help.

"What we've done so far is to make a map of the microwave sky from the data." He pulled out a colored computer printout showing the entire sky in an oval-shaped projection, the standard for such maps. Three areas, covering just about half the map, were brightly colored. To the lower left was a patch of blue, representing the cooler cosmic microwaves in the part of the sky Earth is moving toward. The upper right was red—the part of the sky we're moving away from. And in the center was a lozenge-shaped yellow region: the warm glow of microwaves from the Milky Way.

Then Page pulled out another map. On this one, these prominent, well-understood areas of microwave anisotropy had been electronically subtracted from the data, leaving only the primordial microwaves we would see if the Earth were utterly stationary in the universe and floating far outside the Milky Way deep in intergalactic space. The map was still colorful, but now the colors were spread all over the page, a collection of pointillist dots that added up to no discernible pattern.

It looked like so much static, but only a rigorous mathematical analysis would tell whether it was, or whether Page and his fellow balloonists had finally plumbed down to the level where the underlying structure at the edge of the universe had become detectable, the ultimate test of the cold dark matter model of the universe. Page was deeply immersed in that analysis. The dots, which represented subtle variations in the temperature of the microwave background, could have any of four causes. The first two were subtle contamination from the atmosphere and the galaxy—the same kind that Staggs and Wilkinson had been struggling with in Green Bank. The third was an algorithmic problem—a problem in the software that processes the raw signals into calculations of temperature. And the fourth, said Page, was that the variations were real. "It's possible though not likely that we've actually begun to see the primordial fluctuations at last. We're trying to eliminate the noise," he said. "Right now, our best limit is one point six times ten to the minus fifth. But we should be able to get that down to maybe seven times ten to the minus sixth. If we can do that, we'll know we have something really interesting."

CHAPTER FIVE

WEIGHING THE UNIVERSE

I t is no accident that most of the world's big telescopes are located in the Northern Hemisphere. That is where most of the astronomers and most of the money for building big telescopes come from. Yet a true understanding of the way the universe is put together would be seriously incomplete without intensive monitoring of the southern skies as well. The center of the Milky Way rises high only in the south. So do the nearest stars to our own, the three members of the Alpha Centauri system (the very nearest of which is faint, red Proxima Centauri). So do the nearest galaxies, the Large and Small Magellanic clouds (Andromeda is the closest full-sized galaxy, but the clouds, small, shapeless collections of stars that orbit the Milky Way, are technically considered galaxies as well). So did the only supernova to explode anywhere near Earth since just before the invention of the telescope—the star that exploded in the spring of 1987 was located in the Large Magellanic Cloud. And so, too, does the Great Attractor.

Anyone who wants to make meaningful statements about the cosmos as a whole must sooner or later study the southern as well as the northern skies. Imagine, for example, that a careful survey revealed twice as many galaxies in the southern skies as there are in the northern. That would be the ultimate crisis; it would mean that the universe is lopsided on the largest imaginable scale, and that all existing models of the universe, including the Big Bang, are worthless. (In fact, astronomers are looking for just such an effect. They haven't seen it yet.)

Learning whether the universe leans to one side is only the most extreme example of why astrophysicists need to look in all directions. They also have to know whether structures like galaxy-encrusted bubbles uncovered in redshift surveys are detectable in all directions, or whether they are flukes. A system of voids and bubbles in a part of the sky far away from Huchra and Geller's is convincing evidence that the universe in general looks that way. Recognizing that they can't afford to neglect the southern sky, the institutions that operate observatories began in the early 1970s to build big telescopes below the equator.

The requirements for siting a major observatory are straightforward enough: clear, dry skies; stable airflow, to keep atmospheric turbulence to a minimum; high altitude, to keep telescopes from having to look through too much air; minimal interference from nearby city lights; reasonable access for astronomers and technicians. There are not a lot of places that fit these criteria, and so observatories, like galaxies, tend to be clustered—they have a positive correlation function. One cluster is centered on Tucson, Arizona; another sits on the rim of the crater on top of the inactive volcano Mauna Kea, on the island of Hawaii; another is in Southern California (the Mount Wilson and Palomar observatories there were built before Los Angeles and San Diego were so big and bright).

In the Southern Hemisphere, the richest cluster of observatories sits in the foothills of the Andes, in northern Chile, where some of the clearest, cleanest, and driest skies in the world are found. Three major observatories are located there: the European Southern Observatory on the peak of La Silla, run by a consortium of

European countries; Las Campanas Observatory, operated by the Carnegie Institution; and Cerro Tololo Inter-American Observatory, run, as is Kitt Peak, by the National Optical Astronomy Observatories.

The nearest town to all three observatories is the seaside university town of La Serena, about an hour north by plane of the capital, Santiago. Since each observatory has several good-sized telescopes and since observing runs usually last only a few days, there are always astronomers arriving and departing. During the few minutes between the time a plane arrives from Santiago and the time it turns around and flies back, the parking lot outside La Serena's tiny airport can claim more world-class astronomers than most university departments. On a chilly, cloudy afternoon in October—early spring, in the Southern Hemisphere—I stood waiting for the morning plane with three of the best. All of them were deeply involved with trying to answer questions that bore directly on the theoretical crisis, trying to give theorists some data to work with in understanding why their models were failing.

Bob Kirshner, the Harvard observer who had helped find the Great Void in Boötes, was on his way home from Las Campanas; he had spent several nights working on a new redshift survey deep into space to try and see whether even bigger voids and Great Walls lurked beyond the known structures. John Tonry, from MIT, had spent a comparable time at the four-meter telescope at Cerro Tololo, trying to gauge the distance to relatively nearby galaxies thus creating a scale of megaparsecs others could use to calibrate their maps in ordinary rather than velocity space and, at the same time, to measure the size and age of the universe itself. And J. Anthony Tyson, from AT&T Bell Laboratories, was on his way up to use the telescope Tonry had just vacated.

I had traveled to Chile with Tyson. It had taken us three leisurely days to get from Newark airport, in New Jersey, to La Serena, and in another hour we would be on the mountain—on a project that had consumed much of Tyson's considerable creative energy over the past several years. Like Huchra's and Kirshner's work, this project was in the area of mapmaking as well. But Tyson was only incidentally concerned with where the galaxies are. He was trying instead to chart the location of something that is utterly invisible: the dark matter that pervades the universe.

While specific theories about dark matter—like CDM—may be wrong, the fact of dark matter is inescapable. It's there, herding the stars and gases of the Milky Way and Andromeda and other galaxies so that the flat spirals stay flat and so that all spiral and elliptical galaxies keep from flying apart from their own rotation. It's there on a higher level keeping clusters of galaxies from spinning their individual members off into intergalactic space.

But these are only general descriptions of where the dark matter lies. It surrounds galaxies with invisible halos, but how big are the halos? How far out beyond the visible stars does the halo of invisible stuff extend? Within clusters, how is the dark matter arranged? Bunched tightly right in the center of a cluster of galaxies like a pit in the middle of a peach? Spread smoothly throughout the cluster like a pudding embedded with galactic raisins or clumped independently of the galaxies but intermingled with them, as the popcorn is intermingled with the peanuts in a box of Cracker Jack? And again, how far outside the clusters, if at all, does the dark matter extend?

Tyson's trip to Chile was part of his attempt to find out the answers to these questions. For years, he had been making pilgrimages to some of the biggest telescopes in both hemispheres—the 4-meter at Kitt Peak, the 4-meter at Cerro Tololo, the 3.8-meter at Mauna Kea—whenever he could get the observing time. He was out to create maps of dark matter that could help theorists understand what it is by showing them where it is, and how much.

Unlike John Huchra, Tyson is not formally associated with any observatory, and because he needs the most powerful telescopes to do his observing, he could not easily afford to miss many nights. Other than the run at Cerro Tololo, all the telescope time he would have for the next nine months was one more run here and two at Kitt Peak. The heavy overcast at sea level didn't worry him, since that had no bearing on the conditions at seven thousand feet. But he needed to know what was happening on the mountaintops, so most of his conversation with John Tonry and Bob Kirshner at La Serena's airport was not about galaxies or stars but about clouds. "It was great at Las Campanas," said Kirshner, but that was not necessarily helpful information: it can be crystal clear there and completely overcast at Cerro Tololo. But Tonry's report was good. "We took data every night," he told Tyson. "I don't know if it was photometric, but we'll see." (Photometric means that essentially every photon that arrives at the top of Earth's atmosphere from a given galaxy makes it through to the telescope. This means the data can be calibrated to estimate an object's true brightness.)

From the moment we met in Newark, the weather had been on Tyson's mind. We had met in a departure lounge at the airport before dawn on a Monday morning. Venus still glowed in the east, having swung around the Sun since the run at Mount Hopkins to become a morning star, rising in the early darkness and then slowly fading from view as the Sun chased it into the sky. Jupiter had moved as well; it was now just a few moon-widths away from Venus. Even in smoggy, light-polluted New Jersey, these brightest of starlike objects were hard to miss. But Tyson had missed them, although they must have been right in front of him during the drive in from his home in rural western New Jersey. He isn't much interested in the planets or the constellations unless forced into discussing them by a question from his ten-year-old son, Kris. Even then his wife, Pat Boeshaar, an astronomer at Drew University who specializes in studying cool stars in the Milky Way, often fields the question.

It would take us three days to get to Cerro Tololo for three reasons: because it really is remote from New Jersey; because travel connections within Chile are not as convenient as they are in the United States; and because Tyson was scheduled to give a late-afternoon seminar at the observatory's headquarters in La Serena more than twenty-four hours before his first night on the mountain. The trip to Santiago, where we would stop for a night, involved a nearly three-hour flight from Newark to Miami, then a nine-hour leg to Santiago, which made for plenty of time for him to gear up for the run and also to talk about what he was planning to accomplish.

When I first saw him at Newark, he was staring thoughtfully at some papers in the briefcase lying open on his lap, and after we had boarded the plane, he invited me to see what he had been working on. It was a sheaf full of grimy-looking pictures printed on fax paper—images that were not cosmological but meterological. "These are photos of South America taken by a geosynchronous satellite," he said. The satellite sits in orbit a bit more than twenty-two thousand miles above the Equator. At that altitude, the natural orbital period of a satellite is exactly one day. The satellite thus sits over the same spot on the planet and constantly monitors the cloud cover over one half of the Earth. Its signals are beamed to the ground, and retransmitted twice a day by the navy. They can be picked up by anyone who knows how, and Tyson does. "I'm a radio ham," he explained, "and I have software that captures the signal and transforms it into images." He also receives signals from polar-orbiting Russian satellites, which whirl around the planet at right angles to and much lower than the geosynchronous satellites, passing over a different area on each orbit as the planet turns under them. "There are at least two that pass over the United States once a day," said Tyson, "and you can pick those up directly from the satellites with a scanner." The only difficulty is that drag from the upper atmosphere affects the orbits, so he has to call NASA every few weeks to get the new orbital information.

The result of all these downlinks and uploadings is the set of somewhat murky images he held, which showed patterns of white, gray, and black superimposed on an outline of the South American continent. "Here," he said, pointing to the sheet on top. "This is from last night." The whole of northern Chile was hidden in glaring white. "Those are high clouds. They're the ones we have to worry about, since Cerro Tololo is high enough to be above low-lying clouds." He pulled out another sheet covered with meteorological hieroglyphics. "It looks like a high-pressure system is coming up from Antarctica, though, which could mean that the clouds will clear out by the time we get to the telescope." It seemed somewhat pointless to be able to predict the weather. After all, Tyson had five nights on the four-meter telescope whatever the weather might be and would be able either to observe or not. "It's true," he admitted, "that you get what you get as far as weather is concerned. But you do want to be prepared for different conditions. If the sky is clear, I might run aggressively through the program I have planned, but if it looks as though it will be marginal, I need to be ready to shift gears quickly."

Satellite downlinks are nothing new for Tyson. He has been fiddling with them literally since it became possible to do so. Growing up in Carlsbad, in Southern California, in the 1940s and 1950s, Tyson built radios the way other kids build model cars. When Sputnik was launched in 1957, he found out what frequency the Soviet satellite was broadcasting on and tuned in (he made a tape of the spacecraft's radio beeps for the *Los Angeles Times*). "I remember going up to Mount Palomar in the late 1940s to see the new two-hundred-inch Hale telescope," he recalled, "but I was mostly impressed with all the dials and gears. To hell with the telescope—the control panel was where it was at."

Except for his prematurely and completely gray hair, Tyson still looks more like a kid who cobbles together radios than an eminent astronomer. His face is boyish, and his body is trim and energetic—he walks about 40 percent faster than most people would find comfortable. In fact, he is one of the most talented and creative optical astronomers in the world. He is the man John Huchra calls Tony the Tiger, a tribute not to his fierce demeanor (he is in fact rather shy), but to his observing skills. His dark matter project was typical. Not only was it very difficult but it also was potentially crucial in understanding why the cold dark matter model of the universe had gone so wrong. Tyson's observations might be able to save the model . . . or destroy it completely.

Mapping something that is invisible calls for an unusual amount of inventiveness, and that is Tyson's strong suit. Over his astrophysical career (which is not the same as his scientific career), that inventiveness has been concentrated mainly in two separate tracks, each of which feeds off the other. Tyson is legendary among astrophysicists for his creative use of technology—telescopes, detectors and computers—to make the most sensitive possible observations of the deep universe. At the same time, he finds ways of

J. Anthony Tyson   PHOTO: EILEEN HOHMUTH-LEMONICK

using what the universe has revealed about itself, to him and other observers, to devise entirely new ways of studying it. He is a sort of jujitsu astronomer, using leverage to turn his adversary's strength to his own purposes.

The dark matter maps he was heading to Chile to work on were the result of this sort of strategy: an innovative observing technique that led to a discovery which led to another innovative observing technique which incorporated someone else's discovery—all of it driven by technological advances that Tyson had a hand in. The story is somewhat involved, and it took us from somewhere over the Caribbean to just over the Colombian-Bolivian border for him to explain it to me.

In the late 1970s, Tyson was working at Bell Laboratories, the research and development arm of what was then known as the Bell System. "I had gone to Stanford as an undergraduate," he said. "I started out in electrical engineering, and I lasted about a month. I remember sitting in a course called Motors, and my eyes glazed over. So I switched to philosophy, which I really enjoyed. By my junior year, though, I realized that there wasn't a whole lot of future in it. But I'd been taking physics courses all along and hanging around the physics department quite a bit. When I came over to change my major, the department secretary looked at me and said, 'You'll never regret this.' " Tyson went to graduate school at the University of Wisconsin, where he studied low-temperature physics. "I was fascinated with the behavior of superfluid helium," he said. "When it's cooled to within a few degrees of absolute zero," he explained, "liquid helium starts behaving in peculiar ways, like climbing up and out of containers." Later Tyson became interested in general relativity, and since there was no one at Wisconsin he wanted to study with, he began spending a lot of time at the University of Chicago, where he studied with, among others, Subramayan Chandrasekhar, the Indian theoretical astrophysicist who did pioneering work on black holes and neutron stars.

"When I got my first job, in 1969," he said, "it was at Bell Labs, where they wanted me to set up a laboratory for studying low-temperature physics." He went on to build a device to try and detect gravitational waves passing by the Earth. As part of his general theory of relativity, Einstein had predicted that massive objects accelerated to high speeds—two neutron stars orbiting each other, for example, or a star being swallowed by a black hole—should emit ripples in space and time that propagate through the universe, literally stretching and contracting any object they pass through. The distortions are extraordinarily feeble, and the first gravity-wave detectors, built by Tyson and others back in the early 1970s, consisted of massive blocks of metal attached to extremely sensitive, superconducting (and thus super-cooled) electronic detectors. "By 1973," he says, "we had the world's largest device." It ran until 1984 without success, and physicists are now designing even more sensitive detectors based on finely tuned laser beams rather than chunks of metal (Tyson was asked to testify before Congress in 1991 about whether the $211-million Laser Interferometer Gravity Observatory should be funded; he regretfully told them that he didn't think that the technology was ready for LIGO to go ahead.)

By 1974, though, Tyson's interests began to change once again, and he started thinking about moving further into astrophysics. Bell Labs had always had a strong tradition of supporting pure science, and he was encouraged to do pretty much what he wanted. The presumption was that giving bright scientists the freedom to follow their interests would as often as not lead to products that would be useful to the company. For every Nobel prize earned by a Penzias and Wilson for discovering something useless like the cosmic microwave background there would probably also be a Nobel for a Bardeen, Shockley, and Brittain for inventing the transistor.

Even after the Bell System was broken up by court order in 1984 and the labs were forced to become more product-oriented, there remained vestiges of the old freedom. "I now work on more applied projects than I used to," said Tyson, "but I still find Bell

Labs attractive in that I can get funding for an experiment essentially right away if my director approves of it. And even though we're a corporation, Bell Labs still appoints managers from the ranks of scientists. So when I go to a manager with a project, I can argue it on the scientific merits."

One of Tyson's first astrophysics projects was a study conducted at both Kitt Peak and Cerro Tololo of galaxies that were faint in visible light though bright in radio energy. One night at Kitt Peak, rained out of the evening's work, Tyson sat down to take a closer look at some photographic plates he had brought along. "I had just read a book by Jim Peebles," he said, "in which he talked about galaxies being more clustered than you'd expect, but then admitted that the data only went to eighteenth magnitude, and so weren't really adequate." But the plates Tyson had in his hands went all the way to twenty-fourth magnitude. So he set out to count all of the galaxies on every photographic exposure.

It nearly drove him crazy. There were just too many galaxies, and he was convinced that his counting technique was too imprecise to be reliable. He needed to find some way of counting the galaxies automatically without confusing galaxies with stars on the plate. So he picked up the Bell Labs telephone directory he had brought with him, found the image processing department, and called until he connected with a man named John Jarvis. Jarvis, it turned out, dabbled in low-temperature physics, too, and he was also an amateur astronomer. They talked over the problem for a long time on the phone that night, and when Tyson got back to New Jersey they sat down and created a computer program called FOCAS, for faint object classification and analysis system, which scans an image and classifies every object it finds by brightness and shape.

Armed with FOCAS, Tyson and a series of collaborators began peering ever deeper into space, counting galaxies as they went. One night in 1984, Tyson and Patrick Seitzer, at the time a postdoctoral student at Kitt Peak National Observatory, were taking deep images at Cerro Tololo, and they decided to try taking a risk.

They pointed the telescope at an area of the sky where conventional photographs showed no stars, no galaxies—nothing but blackness. "Neither of us was prepared for what we saw that night," he told me. "We were in shock. We would have been surprised to find one hundred objects in the entire field. Instead, there were at least a thousand smudges of light, a thousand galaxies where we had assumed there was just blackness."

It would have been an inconceivable coincidence if these vanishingly faint galaxies showed up in just the patch of sky Tyson and Seitzer had looked at but nowhere else, and subsequent deep images proved that there was no coincidence at all: wherever he looked, Tyson found that the sky was wallpapered with a hitherto undetected population of faint galaxies. And the deeper he looked into the universe, the more of these faint smudges showed up. "We've gone out to twenty-eighth magnitude," he said, "and the number counts are finally beginning to level off at about a million per square degree of sky. But that's still kind of a problem." The problem, as he explained, is that a million per degree adds up to about forty billion over the entire sky. "There are too many of them simply to be the ancestors of modern galaxies [there are only about twenty billion of these]. No one has yet been able to explain what happened to them."

These ephemeral patches of light, thousands of times fainter than the galaxies Huchra was hunting in the Century Survey, are far too faint for measurements of redshifts. Smearing their light to reveal redshifted spectral lines would make the galaxies disappear entirely. (Once, while Sandra Faber of the University of California, Santa Cruz, and the Seven Samurai was rhapsodizing about the potential power of the ten-meter Keck telescope then under construction atop Mauna Kea in Hawaii, I asked her whether its light-gathering capacity, more than four times that of the MMT, would make it possible to get spectra of Tyson's galaxies. "Well, no, not *those*," she said. When I told him that, Tyson agreed. "The Keck is much too small," he said. "We really need a national twenty-five-meter optical telescope [that is, a telescope with a mirror eighty feet across]. Even then, you'd need all night to take the spectrum of a single one of these galaxies."

Without spectra to reveal their redshifts, it wasn't possible to prove at first that these galaxies were far away. They might be close by, but intrinsically very dim. Tyson believed, though, that they were in fact very distant. The clue was their color: they were unusually blue. "The standard theory up until a few years ago was that to look for primordial galaxies you should look in the infrared because the early galaxies would be dusty. Dust makes the light from a galaxy look redder, and when the red light was redshifted it would move into the infrared. I never really thought much of the idea: without a few generations of stars to produce heavy elements, where would the dust come from? In fact, it looks as though the stars in this galaxy are young and hot, which you'd expect early in the history of the universe. Young stars radiate a lot in the ultraviolet. Redshifting makes them look blue."

By now the plane was ready to land. It was about eight-thirty at night, and we were about twenty minutes late (we'd detoured around a storm system in Bolivia, to Tyson's great interest) when we set down on the runway at Marino Benitez International Airport. There is a routine here just as in Tucson, and it involves taxis as well. The Cerro Tololo administration had assigned a driver to meet us, and Tyson was confident he'd recognize the man from previous trips. With uncanny skill, we managed to be last in line for customs and immigration, but when we finally got through, the driver was still waiting. He spoke no English. Tyson speaks no Spanish. ("I really should learn," he admitted.) But the man's face was familiar, and he already had instructions: take me to a hotel, take Tyson to the observatory's guest house in a residential section of the capital, and then reverse the process the next day in time for the flight to La Serena. There was a slight delay while the driver searched through his car for the antenna he had carefully unscrewed and stowed for safety while parking. ("Hey, I can understand what an antenna means to a man," said Tyson the ham.) Then we were on our way, sharing the dimly lighted two-lane highway with overloaded, swaying trucks and plodding buses. At one point, we passed a horse ambling along with three rustic-looking passengers. We switched to a bigger road and entered the city itself: on our right was a river, on our left and several hundred feet above us was an enormous, brightly lit statue of the Madonna with her arms outstretched. The hill she stood on was invisible, and she seemed to float unsupported in the sky above Santiago. Then we made a quick right turn, across the river, and stopped in front of a hotel that was reassuringly nonexotic. Tyson and the driver drove on.

The plane to La Serena would not be leaving until late in the afternoon, and Tyson suggested a visit to the pre-Columbian museum downtown in the morning. He had tried to go on previous trips, but it had always been closed. The reason, it turned out, was that he had never tried on a Tuesday or a Saturday, the only times the museum is open. But the next morning was Tuesday, and after breakfast he walked from the guest house up to the hotel, through neighborhoods that looked more European than South American. We hopped on the Santiago subway and rode into the city center. It was midmorning, but the streets were filled with people, including schoolchildren in blue-blazered uniforms. As it turned out, in both Santiago and La Serena, there would be no time of day when people weren't crowding the streets, and the crowds got even worse at lunchtime. In the Plaza des Armas, a public square in the middle of the downtown area, soldiers with machine guns patrolled the sidewalks, their fingers on the triggers of Uzis. It was a reminder that General Pinochet still ran the country's security apparatus, although he had stepped aside as president. People no longer disappeared to be found years later, if their families were lucky, in secret mass graves. But it was also clear, despite antigovernment slogans and posters on some of the buildings, that the military junta that toppled Salvador Allende in 1973 had not disappeared entirely.

Tyson is utterly at home navigating through the skies, but in downtown Santiago he found to his surprise that he was heading north rather than south. "I imagine you've never been shown around by an absent-minded astronomer," he said. We reversed direction, consulted a map, and made our way to the museum. Tyson was captivated by the artifacts. "I find these things amazing," he said. He pointed at a two-spouted water pitcher shaped like a grotesque dog. "I'd love to bring one of these along to the theoretical tea at Bell Labs." He was silent for the rest of the visit, bending over to study each figurine and weaving carefully. Finally he straightened up and said, "How about lunch?" He could only vouch for one restaurant in town, on the square where the Moneda, the presidential palace, is also located. It was appalling to think that air force jets had bombed the palace in 1973, not only because that is not the way power should change hands in a country with a long history of democracy but also because the palace is in the middle of a modern city, an area where thousands of people live and work.

After lunch, we found our way to the subway without incident and headed back to the guest house to wait for the driver. Cerro Tololo shares its Santiago guest house with the other observatories, and like the airport at La Serena, it is a place where chance encounters between astronomers can happen. As we walked in, we saw an older astronomer, dressed like a farmer in khaki pants, a plaid shirt, and red suspenders, sitting on a couch. "Well, well," said Tyson, looking surprised and pleased, "hello, old man. What are you doing here?"

His name was Ray Weymann; he worked at Caltech, and he was heading up to Las Campanas to peer at quasars. He was also part of a team of observers who had discovered a cosmic phenomenon that both contradicted a prediction made by Einstein and gave astrophysicists a remarkably useful tool for probing the universe—the same tool that Tyson was using to map the dark matter. The tool is, in essence, a mirage on a cosmic scale.

In 1979, Weymann and his collaborators were looking at quasars, intensely bright but very distant points of light that had been discovered nearly two decades earlier. At first, no one knew what the objects were. They looked for all the world like dim, bluish stars. But nobody could figure out what elements their spectral lines corresponded with. Finally, Maarten Schmidt at Caltech realized that they were ordinary lines, but redshifted so unexpectedly far that astronomers had failed to pick up on them. The first, 3C273, had a redshift of .16, and the second, 3C48 was up at .37. These "stars" were more distant than all but the most distant galaxies, and judging from their light, they were inherently about one hundred times brighter than an entire galaxy.

Since then, thousands more quasars have been found, and their redshifts have constantly increased. The record holder has a redshift of just under five, which means it is more than 85 percent of the way back to the edge of the universe. Astronomers think, but cannot be sure, that the power behind the quasars is black holes, objects at the centers of galaxies that are at least as massive as a million stars, but collapsed into an area so small that their surface gravity won't even let light escape. The idea is that stars and gas falling into this gravitational bottomless pit are squeezed as they spiral in, heating up to incandescence as they go. Black holes paradoxically, are probably the brightest objects in the universe (smaller holes probably exist as well, even inside the Milky Way, but they don't shine so brightly).

Weymann and his co-observers were doing the nuts and bolts work of quasar hunters, recording the objects' spectra to try and understand their composition and behavior, when they found two quasars just a few arcseconds apart in the sky. That seemed awfully close. Moreover, the quasars' spectra and redshifts were essentially identical. That seemed wildly improbable. Identical redshifts meant they were at identical distances from Earth, and thus were lying very close to each other in space, something that had never been seen before. Moreover, quasar spectra are something like fingerprints or snowflakes, all pretty much alike in overall structure but very different in detail.

The utter improbability of having two identical quasars in the first place and having them sitting right next to each other in the second led Weymann and his group to conclude that there weren't two quasars at all. There was one: the apparent doubling was an optical illusion, a mirage caused by the presence somewhere in the foreground of a massive object, perhaps a galaxy. According to the general theory of relativity, a massive object literally warps the space around it, causing light rays to bend instead of going straight. We can't see the bending; all we see is light arriving at Earth from a direction other than the direction in which it started out. The practical result is that an object—a star or a quasar— whose light has come close to a concentration of mass will appear to be slightly out of position. The prediction, along with Einstein's theory, was vindicated during a solar eclipse in 1919, when stars very close to the blacked-out Sun were seemingly shifted by just the amount Einstein had calculated.

In 1936, someone asked Einstein what would happen if one star passed directly in front of a far more distant star, and he obligingly responded with a short paper in the journal *Science*. Having performed what he called a "modest calculation," Einstein concluded that the star in front would magnify and distort the image of the star in back. If the two were perfectly aligned, he explained, the background star should look not like a dot of light but like a tiny, perfect circle. But, he added, the phenomenon would never be seen, since the chance of one star passing in front of another would be vanishingly small, and the ring would be too minute to see.

That is true enough, but Einstein didn't know about quasars, which lie beyond all but the most distant galaxies. Since galaxies are spread out, not pointlike, it is not unlikely that one should occasionally get in front of a quasar. That is just what Weymann had seen: the first example of a true gravitational lens. Although Einstein did not get into the details in his two-page paper, it turns out that the actual distortion of a background object by lensing depends on many factors: the shape and mass distribution of the body doing the lensing, the shape of the background object being lensed, the alignment of the two, and their relative distances. Weymann's lens turned a single quasar into an apparent pair of twins, but gravitational lens systems discovered since then have had triple quasar images, quadruples (John Huchra found one of these), and even rings in which not the quasar itself but a jet of matter shooting from its core appears on the sky as a minute, nearly perfect circle. Sometimes the background object is not even a quasar; it's a distant galaxy, and when that's the case the lensed image often ends up looking like a fragment of a circle, an arc of light. To complicate matters, the light-bending mass in the foreground is, in many cases, not just a single galaxy, but a cluster of galaxies, both more spread out and more massive than a galaxy and thus even more likely to interrupt and distort the light coming from beyond.

Tyson had been intrigued with the theory of gravitational lensing ever since he read a paper on the topic in 1973, and when Weymann and company actually found one, he made a creative leap. Mightn't it be possible, he wondered, to look at a distorted background object and deduce from the distortion what sort of foreground object had caused it? He first tried the technique in 1979, seeing whether background galaxies were lensed at all by those in the foreground. "We basically didn't find an effect," he told me in the taxi to the airport. (Weymann and another astronomer were headed up to La Serena on the same flight we were taking. They were going to look for more quasars, and after a few minutes of comparing research notes we got into our taxi and Weymann into his. Our drivers jockeyed for position down the boulevards of Santiago all the way to the airport.)

Failure to see anything was in itself significant. By finding nothing, Tyson had proven that the dark matter halos around individual galaxies didn't extend indefinitely. They had to be less than one hundred kiloparsecs, or about three hundred thousand light-years in radius; otherwise, they would have caused the background galaxies to look warped. (The Milky Way has a radius of about fifty thousand light-years.)

Once he had learned that the halos of individual galaxies are limited in extent, though, there was not much more he could do with them. Their relative compactness meant that a background galaxy would have to be very close on the sky to one in the foreground for lensing to take place, and that, considering the overall numbers of distant galaxies in the universe, appeared to be statistically improbable. But that was before his discovery of the faint blue background population. "When we found this new population of faint blue galaxies," he said, "what I realized was that there were enough of these background galaxies to let us trace the distribution of mass in the foreground." He also realized that the blue galaxies were far enough away so that he could use entire clusters rather than individual galaxies as his lenses. Because clusters look bigger on the sky than individual galaxies, and because they contain more dark matter in percentage of total weight, they should be doubly useful as gravitational lenses.

The distortions he was looking for were subtle, though. While Tyson's deepest images reveal perhaps one hundred faint blue galaxies in and around the average cluster, they are small enough, and the cluster's mass spread out enough, that the lensing effect on each one would be almost imperceptible. The galaxies should be stretched into arcs, fragments of circles centered on the core of the cluster, but the stretching would be minimal. Because the faint galaxies appear as little more than blobs of light, and because galaxies usually look elongated in one direction or another anyway, no single blue galaxy could tell him much of anything. Only by analyzing the images with FOCAS and then studying the statistics of the blue galaxies' shapes—how many are elongated by how much and in what directions, compared with what you'd expect to see in an unlensed system—would he be able to say anything at all meaningful about the distribution of mass in the cluster.

"The first time we tried to do it," he said as we spotted Weymann's taxi just moving up on the left, "we looked at twelve hundred background galaxies to see if they were lensed, and they weren't." The result wasn't really definitive, though, because of the medium he was working in. "We were still using photographic plates at that time, and they're really horrendous. Their quantum efficiency is about one percent, which means that only one out of every hundred photons that falls on them is detected." Not only that, but with photographic plates, an astronomer is using a new detector for each exposure. No matter how carefully it's made, each plate is just slightly different from the next. The density of the photographic emulsion can vary slightly or be a hair different in its chemistry or it can be developed at a fractionally different temperature. That puts a limit on how accurately the images on a given plate represent what's in the sky.

"Right now," said Tyson, "we're doing a test, trying to push some Kodak plates as far as we can by exposing them for five hours apiece. I think we'll be able to go to twenty-fifth magnitude with them. But considering the trouble we're going to, and the fact that with CCDs [charge-coupled devices] we can go five magnitudes fainter, it does seem a little bit crazy. In a way, it's lucky that we started out using plates, though. They were so bad they forced us to develop the image analysis software we're now using."

CCDs are electronic light detectors that have transformed astronomy over the past decade and a half. Their quantum efficiency is close to 100 percent. They can record information one hundred times faster than photographic plates, and they are reusable. A CCD is a sort of silicon chip. Whenever an incoming photon whacks into its surface, the photon dislodges an electron and sends it crashing into the next layer, which is laced with a gridwork of current-conducting channels that mark off squares. When an electron falls into one of the squares, an applied voltage keeps it there, and as more photons fly in, more electrons fall, each in the square right underneath the place where its photon fell. Some boxes end up with a lot of electrons (lots of light hit right above these), some have a few, and some none at all. The distribution is exactly the same as the distribution of light that hit the CCD from deep space (or from the family dog: CCDs have become common and cheap enough that they're now used in video camcorders).

Finally, when the exposure is done the computer that controls the CCD begins varying the voltage in the wires, setting up waves of electric potential that ripple across the chip. The electrons surf on the waves, one row of grid squares per wave, until they reach the edge and then they fall off onto a wire that carries them away for electronic counting. The computer counts the electrons one row at a time and records the count in its memory. When the last row has been read, the computer can reconstruct the entire image on a video screen or store it for later viewing. Once the image has been stored, it can be added to later on: point the telescope at a particular galaxy for an hour each night over a hundred nights, and you get the equivalent of a one-hundred-hour exposure.

CCDs do have a drawback: they're tiny. The photographic plates used in astronomy measure up to twenty inches square, and they can take in several degrees of sky at once. The CCD Tyson would be using on Cerro Tololo was barely an inch square. It's hard to grow very pure silicon crystals of any great size, and while it would in theory be possible to cobble together the equivalent of one big CCD from many little ones, the number of electrical connections and the possibility of one of them failing makes it impractical.

But Tyson the satellite downlinker is an inveterate tinkerer, and he was working on the small-CCD problem in New Jersey. Along with a handful of other astronomers, notably James Gunn at Princeton, Tyson is an exception to David Wilkinson's distinction between astronomers and physicists. He does build his own instruments. Along with several colleagues, including Gary Bernstein, at the University of Arizona, and people at Tektronix, where the chips are made, he was in the final stages of building a CCD camera whose surface would measure six inches on the diagonal. It would take in a square degree of sky in one shot, enough to capture most of a cluster in a single image. Hitherto he had been limited to just the central regions. "A few weeks ago," he said, "with Gary's help I finally got the electronics working. When I saw how low the instrument noise was, I couldn't believe it. When you look at an image on the screen, you have to hold on to your desk, because you're afraid you're going to fall off the spacecraft. And that was with hand-wired circuits. The printed circuit board that we'll actually use in the field will be even better."

Although the flight to La Serena wouldn't leave for an hour and a half, Tyson insisted on checking in right away. He wanted to be sure of getting seats on the right side of the plane. "There's a significant anisotropy in the view," he said. Then we settled down to wait, sitting on the grass outside the terminal. "I was resting out here once," said Tyson, "and fell asleep. I woke up to my name being announced at maximum volume, booming across the parking lot." If we hadn't been there, Weymann might well have found himself in the same predicament. He had taken the overnight flight from the United States and hadn't gotten much sleep (he had only been at the guest house for a few hours when we arrived). But he hadn't seen Tyson for a long time and couldn't resist talking shop.

Weymann had continued to work on quasars rather than lenses. He was using them to try and probe the giant clouds of intergalactic hydrogen gas that lie between the quasars and Earth. These clouds might qualify as dark matter, since they can't be seen directly, but they can be positively identified because they absorb some of the light of objects behind them. Take a spectrum of a quasar, and about 20 percent of the time, it will have what is known as a damped Lyman-alpha spectral line superimposed on it. Sometimes there will be more than one at different locations indicating clouds at different redshifts. No one really knows how many clouds there are, how they're distributed, or what part they play in overall structure. No one knows for sure whether the clouds are associated with galaxies or whether they're out between the galaxies.

"You know," said Tyson, "one thought has occurred to me. At a million per square degree, the faint blue galaxies cover about twenty percent of the night sky. It's also true that about twenty percent of quasars have broad absorption line systems, which means that their light is passing through a thick layer of gas on the way to us." He raised his eyebrows suggestively; perhaps the clouds were simply the gas in and around the faint galaxies.

Weymann thought the hypothesis was interesting but far from convincing. He had another question about the blue galaxies. Just as in the case with the nearby galaxies Huchra and Kirshner were charting, there are patches where there seem to be far fewer of them than average, and he wanted to know what Tyson thought they meant. "It's tempting to think they're being obscured by dust," he said. The notion of voids and bubbles so early in the history of the universe would be one more challenge to the CDM model. "So you go look at the edges of the patches to see whether there are redder galaxies there, galaxies that are partially obscured. We don't see it," he said. "So what's your explanation?" asked Weymann. "Structure." He was suggesting, in other words, that the large-scale clustering of galaxies, which seems to be uncomfortably widespread in the modern universe, might have been around early on as well.

When we finally took off, I could see why Tyson had insisted on the right side. Chile is an extraordinarily shaped country, 2,650 miles long and an average of only a few score miles wide. ("Mathematically," he said, "it's one-dimensional.") Along its entire length, straddling the border with Argentina, are the snow-covered peaks of the high Andes. From where we sat we had an uninterrupted view of the mountain range, stretching beyond the edge of the world both north and south. Aconcagua, an extinct volcano, towered above them all. ("A student of mine once hiked to the summit; it took him three weeks.") The flight would last only forty-five minutes, but the attendants managed to serve lunch anyway. As a tray was placed in front of me, Tyson whispered some advice: "Don't eat it. You have to be careful about food in Chile," he said. "I've gotten sick every time. Last time it didn't hit me until I was in Miami on the way back, and then I was really in bad shape—fever and everything. The doctor told me it was either typhoid or cholera." I asked him how he might have gotten ill. "Eating raw shellfish," he said. "I'm too adventurous for my own good." (On vacation he goes to places that are really exotic—Nepal, and last summer, the island of Vanuatu, in the South

Pacific. "It was great. Every evening at sunset the beach would be covered with these tiny, incredibly poisonous snakes slithering down to the ocean.")

Instead of eating, Tyson got out a piece of paper he had been working on ever since Newark; on it were neat rows and columns filled with numbers and cryptic notes. "This is what I call a train schedule," he said. He explained that it was a chart listing the proposed viewing schedule for each night of the run, with day and time on the left and the objects to be studied, along with information about their celestial coordinates and what filters he would be looking through, on the right.

"We're going to start with a candidate we think might be a rich cluster of galaxies," he said. "Then we'll move to a field I've been pounding away at for six years now. We have about thirty hours of integration so far. This one is particularly interesting: it's the field with the binary quasar 2345 + 007." It is actually a single quasar, lensed, but in this case the split between the two images is unusually large—seven or eight arcseconds. Until recently, no one had found any evidence for what could be doing the lensing: there were no galaxies in the foreground. What they did see, though, was evidence of matching absorption lines in both quasars at a redshift of about 1.5. There is some sort of gas lying in front of both quasar images, superimposing its own spectral lines on the quasars' light. "With these deep images," said Tyson, "we've finally been able to pick up what looks like a cluster of our faint blue galaxies, which could certainly be at just that redshift. I'm hoping to look at this field for an hour each night." If Tyson is right and there is a cluster there, it is the most distant cluster of galaxies ever seen and another powerful clue to how the universe has evolved. Clusters should happen late in the life of the universe, and this early one would put yet another tight constraint on any viable theory of cosmic evolution. "We only know of a few clusters as far away as a redshift of point nine," he said. "We only know about thirty or so above a redshift of point two."

Then he pointed out the window at an almost imperceptible

white dot on top of one of the dozens of snowless smaller mountains between us and the high Andes. "That's Cerro Tololo," he said. "You're seeing the dome on the four-meter." Then the plane veered to the left and plunged down through a thick layer of low-level clouds to land at the airport. It was a thin turnout of astronomers that afternoon, and the standard observatory-arranged taxi took us into town, first dropping me at my hotel and then taking Tyson up the hill to the observatory headquarters and his dormitory room. "Let's meet at the headquarters building tomorrow morning," he said. "I have a lot of people to talk to and some homework to do, but I'll be easy to find. You can take a taxi if you like, but I'd walk if I were you. It'll take twenty minutes or so."

It probably would have taken him twenty minutes, but for a normal person it was more like forty. La Serena stretches from the Pacific up onto a plateau. The university, and the AURA compound (the Association of Universities for Research in Astronomy manages the National Optical Astronomy Observatories, of which Cerro Tololo is one, for the National Science Foundation) are near the top. The town and the university are unmistakably Latin American. The blank-facaded houses, painted pink, white, or pale green, face onto interior courtyards; public squares are filled with uniformed schoolchildren playing and young couples strolling; storefronts open onto the street, staying open for business late into the evening but closing for several hours in midafternoon. But once inside the eucalyptus-shaded compound, you might as well be in the United States. The buildings are standard research-park modern. Even the electric current runs at 110 volts, like home but incompatible with the 220 volts the rest of Chile uses.

I walked into the reception area and ran into Tyson at once. He had collared Mark Phillips, a tall, skinny, bearded astronomer who was acting director of the observatory while the director, Bob Williams, was on leave in Germany, and was giving him the latest news about the new CCD camera. When he was done with Phillips, he would have dearly loved to go talk with Tom Ingerson, one of the local instrument specialists, about putting a satellite weather downlink in the headquarters building and about the camera and how to mount it on the telescope. But he had to do a little more homework before we went up the mountain the next afternoon.

Like every major astronomical installation, Cerro Tololo has in its library a complete set of copies of the Palomar Sky Survey, thousands of photographs taken in the 1950s, that cover the entire sky. Tyson needed to double-check the position of the first cluster we would be looking at. The cluster had been studied by another group of astronomers several years ago, and he consulted a copy of their paper to get the celestial coordinates. "Okay," he said, "we want to look at 2213 minus 36." (The first number is an object's right ascension, that is, its celestial longitude. The second is its declination—the latitude. The only name many astronomical objects have is just these two numbers.) He went to a metal filing cabinet eight feet across and four feet high, scanned the labels on the flat, wide drawers, and finally pulled one drawer out. He leafed through several sheets and pulled out a photograph twenty inches square. It was white, but covered with tiny black dots. Nature has slipped up, making the human eye more sensitive to black stars and galaxies against a white sky than the reverse, and astronomical research photos are generally printed as negatives.

The photo came with a clear plastic overlay on which specific galaxies and stars and their coordinates were marked. "Okay," he said, comparing the photograph with a muddy Xerox of the published paper. In the paper, at least, the tiny section of sky was blown up to legible proportions. "I see these three stars here . . . there it is. That fuzzy place right there." He pointed to an almost invisible set of marks that spanned perhaps a quarter inch. It was now clear why an observatory library was equipped with microscopes.

Now that he had located the cluster, Tyson began checking for bright stars nearby. "I don't want to point into the headlights," he explained. The field of view of the CCD is only a fraction of what the telescope itself can see, but he explained that even outside the CCD's range a star could cause trouble. "If the star is close by," he said, "its light is still hitting the mirror somewhere and bouncing around in the prime focus camera, where it could affect the data."

Tyson came up with this particular cluster, as he does with many of his candidates, in a search through the astronomical literature. He is especially interested in clusters that are "dynamically relaxed"—clusters whose members have completed the process of falling together under their gravity and have entered into the complex dance of their mutual orbits. These have the densest cores and thus are the most powerful lenses. The visible compactness of a cluster is a good indication that it is in fact relaxed. Astronomers measure the speeds of individual galaxies by seeing how much their redshifts differ from the cluster average, but if the cluster is not yet relaxed, astronomers can confuse the speed of the initial falling-in with the orbital speed and get the mass wrong. Another clue that a cluster might have fast-moving galaxies is the presence of X-rays. "X-rays are produced by hot gas," he said, "and you have to ask yourself why a cluster would have hot gas in it. One good answer is galaxies whizzing through it and heating it up.

"By now we've done twenty-eight clusters to about four hours each of integration. What we find is that if the clusters are rich and compact, we always get a big distortion of the background objects, and that it's easy to map the dark matter. When the clusters have not collapsed, there is less distortion. But at a minimum, this technique gives you M over R, the amount of mass within a given radius. And if you want to know what the dark matter is made of, it's important to know how it's distributed. For example, suppose dark matter is neutrinos with a small mass, despite what the simulations have told us. Then, because of the way neutrinos behave, they wouldn't be able to concentrate tightly in the cores of clusters [this is just an extension of the argument Gunn and Tremaine used against neutrinos making up the dark halos of galaxies]. Since our maps have shown us that dark matter does concentrate that way, either the mass of the neutrino is greater than ten electron volts or the dark matter is something else. We've also seen that the dark matter seems to clump where the visible matter does; they share a center of mass. And in clusters where you have two centers, two massive galaxies near the core, you find two centers of dark matter as well, although not distributed exactly the way the luminous matter is. The dark matter clumps are less compact.

"We're counting up dark matter wherever we can find it. So far, the history of dark matter detection has been parallel to the history of our understanding about stellar systems, where we first discovered the stars and only later found the stuff around and between them. Now that we know how much dark matter is in the galaxies, we can start looking in between. One experiment we've been doing, and which we'll take data on during this run, is to look at clusters that are quite nearby to see whether there's more structure in the distribution of lensed background galaxies than we see now. If the overall distribution of dark matter in a cluster is pretty smooth, then a nearby cluster will look too spread out to do much lensing. But if there is substructure—lumps within lumps—it will be able to lens. We've already looked at the Coma cluster and didn't see any blue arcs. So either the dark matter clusters on larger scales, or Coma is unusual."

Another project involves looking in places where there are neither clusters nor individual galaxies in the foreground. According to the cold dark matter model, dark matter should come in clumps of all sizes, with galaxies and clusters forming only where the clumps are densest, just as snow will sometimes dust just the highest peaks of a mountain range while leaving lower peaks untouched. "If we see lensing where there's no visible matter, that could be really exciting," Tyson said.

The images can also be used to study phenomena other than dark matter. "You always want to use the data in as many ways as possible," he said. "Since we have so many exposures of a single field over so much time, we see lots of moving objects. So we're looking for asteroids or comets out beyond Neptune, where they've rarely been spotted. Gary Bernstein and Raja Guhathakurta have a software program that searches for them. And we're also using the images to search for supernovas. In any given image, we have four or five thousand galaxies. On the average, that means that one should have a supernova. No one has ever seen supernovas at these distances. What we're looking for is evidence that a galaxy has changed significantly in brightness over a few weeks or a few months. We've actually spotted some of these, although it's too soon to say whether they're supernovas. They might be active galactic nuclei (AGNs) instead, for example. So we go back to the field a third time. If they're AGNs, they will flicker on and off, but if they're supernovas they'll just brighten once."

He went back to his photo. "I must be hallucinating," he said. "I think I'm seeing arcs right here on the Palomar survey plate." This is quite possible; if a cluster is close enough to Earth, then it's capable of lensing not just the faint blue galaxies but also those that are relatively nearby—close enough to be visible even in conventional photographs. If that's true in this case, said Tyson, "then this might be a candidate for one of those nearby clusters we can use to test the idea that dark matter clumps on small scales." He sat up straight. "Okay, I believe it. This cluster really is worth looking at."

His homework done, Tyson went off to talk engineering with Tom Ingerson for a half hour, and then it was time for lunch. "Your hotel is one of the few places in town you can get a good meal at a reasonable price," he told me, and we walked back through the town at an exhausting Tyson-like pace to get there. We were seated by a waiter Tyson has run into on other trips, and he wanted to try out a theory on me. "Don't you recognize him?" Tyson asked. "He's a dead ringer for John Wheeler." Wheeler, an emeritus professor at Princeton, is one of the most eminent physicists alive, having made fundamental contributions to the fields of quantum theory and gravity. Among them is the concept of the "wormhole," a tunnel through the fabric of spacetime that in principle could provide a shortcut from one part of the universe to another. Maybe the waiter really was Wheeler, picking up extra money in retirement by traveling through a wormhole and waiting on tables at the Hotel Francisco de Aguirre in La Serena, Chile.

As we walked back up the hill to headquarters, where Tyson would be briefing the in-country astronomers and any visitors who happened to be down from the mountains about his work, he talked about the growing discrepancies between theory and observation. "I think the universe is going to turn out to be a lot more complicated than our simple models," he said. "Even star formation is complicated. When you take the spectrum of a cool star, you see all sorts of junk, including steam. If we can't predict the weather a month from now on Earth, how can we understand these stars? And our knowledge of the stars is a lot more advanced than our understanding of cosmology. Dark matter will turn out to be an important part of the story, but not necessarily all of it, and even dark matter might turn out to be messy."

The cold dark matter model presumes that dark matter is made up of a type of elementary particles that have never been observed, despite a number of laboratory searches. That is a little too speculative for many observers, and so they look for more ordinary stuff to make it out of. "Suppose the dark matter were made up of black holes," said Tyson. "That's been one of my favorite ideas since 1973, when I read a paper by Bill Press and Jim Gunn titled "Limits on the Cosmological Density of Compact Objects." (This was the paper that originally got him interested in gravitational lensing.) "They said suppose you have black holes of ten to the eleventh or ten to the twelfth solar masses, enough of them to close the universe [that is, to halt the general expansion with their gravity]. Then you'd see them lensing background objects. If they're much smaller than that, you get to a point where you need so many of them that the chances are one will be close enough to disturb nearby stars. If they were somewhere in the middle, say a billion solar masses, then there would be about a hundred of them in the galactic halo. A fair number will plunge through the disk from time to time, colliding with things like protostellar clouds and blowing them to smithereens. You'd never get star formation.

"So maybe there's a very narrow range of sizes of black holes you could have if they're the dark matter, but that presumes there is only one form of dark matter, and Nature is probably more perverse than that. One other form I like is stars that haven't quite made it, stars that are not quite massive enough to have initiated fusion at their cores. My wife and I have actually been searching for these. Again, if you assume that all dark matter is in this form, with stars of about eight one-hundredths of a solar mass, you'd expect there would be so many of them that you'd see some nearby. Another point: if there are that many stars that didn't make it, there should be enormous numbers that did make it, but just barely. We've done searches with infrared detectors, though, and we haven't seen any. So it's impossible to put all dark matter into subluminous stars. There really are problems with every kind of baryonic matter you can come up with."

By four o'clock, it was standing room only in the conference room at CTIO (Cerro Tololo Inter-American Observatory) headquarters. The room was filled with staff astronomers, visiting observers on their way up to or down from the three observatories (several observers from the European Southern Observatory had been interested enough to make the two-hour drive down from the mountain to hear Tyson. They would head back when the talk was over.) Bob Kirshner was there, just down that afternoon from Las Campanas, and he couldn't resist breaking into the talk from time to time to make irreverent comments. If Kirshner is the David Letterman of astronomy, Tyson is more like Dick Cavett: his sense of humor is utterly dry ("When we get up to the mountain, you'll want to check your room for tarantulas and scorpions," he had told me earlier, with a perfectly straight face. "And watch out for the anaconda." I presumed the former were possible, but the nearest anaconda was across the Andes, easily a thousand miles away.)

Tyson started by posing the three "big questions" about dark matter. "First, where is it; second, how much of it is there? Knowing these two would let you go from first principles and do cosmology. But many of us also want to know the answer to another question: what's it made of? Knowing the last might help us rule out some of the models for structure growth that have been proposed."

Covering the basic ideas quickly, he went on to show a chart of galaxy counts as a function of faintness. As the luminosity grew lower and lower, the number of galaxies rose until at the dim end of the chart were the counts for the faint blue galaxies, hovering at the million-per-square-degree level. "You might assume that these faint blue galaxies are really small and relatively close by," said Tyson, "but the fact that they are lensed shows that they're not—they're clearly behind the clusters."

He acknowledged that the blue galaxies are too faint for taking conventional redshifts. "I can say a little bit about the redshifts, though," he said. "You can set a lower limit," he said, "by seeing how much the blue galaxies are lensed by clusters of galaxies in the foreground." According to the geometry of lensing, if most of the galaxies in the background are distorted it means that they lie at least twice as far away as the lensing object. "We have a whole collection of lenses we can use," said Tyson. "We have some nearby, at a redshift of point one or point two, and we have several at a redshift of point five. They've all been studied, and we think we understand them well. We found a few years ago that for the lenses at point one and point two, virtually all of the faint blue galaxies were lensed, which means that the blue galaxies are at a redshift of at least point four; in fact, we've been able to measure a redshift for several of the brightest, and they range from point seven up to two. The real progress has been in recent years, looking at the clusters at a redshift of point five. In these cases, the majority of blue galaxies in the background are still lensed, but not as many as with the nearer clusters."

The lensing-versus-cluster-distance statistics give a lower limit for redshifts of about .4. To set an upper limit, Tyson and two colleagues took some deep exposures a few years ago through a filter that let through radiation right on the border between visible and ultraviolet light. If any of the faint blue galaxies are at a redshift of more than three or so, an especially dim part of their spectrum would have been redshifted into this band; the filter should then have made them vanish completely. "Only one did disappear," said Tyson. "So a picture is emerging of a faint blue population that lies between a redshift of point four and three, with many at a redshift of point eight or more. This is consistent with the assumption we're making that these galaxies were formed at about the same time as many of the quasars."

Then he showed a photograph, a composite image of a galactic cluster built up from many exposures through three different color filters. The filters, like the pigments in color film, allowed Tyson to reproduce the true colors of the galaxies, near and far. At the center of the image were several bright, reddish blobs of light—the galaxies that made up the cluster. In and around them were the faint blue galaxies. Some looked a little bit elongated, but several were clearly arclike, stretched into single parentheses like no ordinary galaxy could possibly be. "Ah," said Kirshner, grinning and pointing to the most prominent. "That must be the Arc de Triomphe." Tyson winced.

Finally he put up his most dramatic slide: a computer-generated image of the same chunk of space he had just shown, but this time showing the distribution of mass, not light, and thus mapping out the dark matter. It looked nothing like the previous image, except for the fact that both had most of their action going on in the center. There was no evidence at all of galaxies, just three overlapping, roughly circular blobs. The dark matter was lumped like three fused snowballs, right in the center of the cluster, an invisible knot whose powerful gravitational pull was keeping the visible galaxies from flying off into space, and was at the same time warping the appearance of the blue galaxies far beyond. The room was silent for a moment, and then Kirshner guffawed.

Tyson was too quick for him this time. "Yes, I know," he said. "It *does* kind of look like Mickey Mouse." He went on. "It's clear that the luminous matter and the nonluminous matter know about each other. But the dark matter is distributed more smoothly. In addition to the distribution, we can also calculate the overall mass of the cluster. So far, in the clusters we've looked at we're getting two to three hundred times as much mass as the light would indicate."

When the talk was over, Tyson was mobbed by the other astronomers, asking him questions about his observing technique, about the image analysis software, and about ideas they had on what he might want to look at next. Finally, after patiently dealing with each one, he went off to relax for an hour. The astronomers and staff from CTIO and ESO (whose headquarters is just downhill from the AURA compound) were going to face off in a volleyball game, but Mark Phillips had invited everyone to a party that evening, and Tyson needed a break.

Phillips lived in an enclave of modern houses a few miles south of town along the Pan American Highway; it was indistinguishable from what you'd see in an upper-middle-class California suburb. The conversation that evening was about astrophysics, of course, but it also turned to the peculiar problems of doing astronomy in Chile. Tyson introduced me to a short, white-haired Chilean named Victor Blanco, who was the director of CTIO in the 1970s, and the stories he told in between bites of cheese and empanadas made it clear that diplomacy is a major part of the job. "When Allende became president," he said, "I thought it was a good idea to go meet him and tell him about what we do up here, and why it's important. So I went down to Santiago and tried to get an appointment. The best I could do was to speak with one of his aides, but at the last minute the man said, 'The president has decided he would like to see you.' We went through a door, and there was Allende . . . and perhaps thirty photographers. The next day the pictures were all over the newspapers, with the headline EVERYONE FROM PEASANTS TO ASTRONOMERS WANTS TO MEET AL-LENDE.

"A few years later, we were having trouble with Argentina, and it looked like we were going to have a war. I could just imagine their air force popping over the Andes. The first thing they would see are our nice white domes on Cerro Tololo, which would be beautiful targets. I flew to Buenos Aires and asked the astronomers over there to explain to their government that we weren't some secret military installation. They did, and I was assured that if it came to war, the domes would be safe."

Tyson has had to deal with the vagaries of Chilean politics as well. "I first came down here in the last days of the Allende regime," said Tyson, "and the government was really getting touchy, with good reason I admit. We were bringing in a new detector, and the soldiers at the airport insisted on looking inside the dewar. They were sure there was something inside. There was: a vacuum. We had a terrible time running around in La Serena looking for a vacuum pump to fix it." Vera Rubin later told me that she, too, had had problems in Allende's last days. She had been observing at the four-meter, and it became evident that foreign nationals should get out as soon as possible. Her husband, who had been along for the trip, drove north and escaped over the Peruvian frontier. She stayed a few days longer, taking just a little more data, and then headed for Santiago. "The planes weren't running by this time," she said, "and so I drove with an American astronomer and his family who were leaving as well. As we got closer to the capital, the checkpoints on the road became more frequent and the soldiers were carrying bigger and bigger guns. Finally, at the airport, we were told there weren't enough seats for us all to leave, so I volunteered to hold one of the children on my lap." Life was periodically hazardous during the reign of Pinochet as well. La Serena was a hotbed of political opposition to the regime, and Tyson recalled a night when he and several other astronomers were pulled over by the police. A bomb had gone off at the local power station; a curfew had been declared and the astronomers had been caught violating it. "We should have realized something was up when there were no cars or people anywhere on the street," he admitted.

Two decades later, things had settled down considerably. The shuttle up to the mountain left at a civilized eleven-thirty the next morning, but I had to hike up through the town for what seemed to be the twentieth time to get it. On the way out of town, we stopped at the airport not only to drop off a staff astronomer who was flying down to Santiago but also to pick up one of Tyson's collaborators. Raja Guhathakurta was a postdoctoral fellow at the Institute for Advanced Study, and he was working with Tyson and Gary Bernstein on the dark matter project. He would have come down earlier, but he had been observing at Kitt Peak until just a few days before, and he had a trip that was even more involved than ours: as soon as he got off the mountain in Arizona he was scheduled to fly from Tucson to Miami, transfer to a flight to Santiago, and then, because there wouldn't be a plane to La Serena until the next afternoon, he was going to transfer to an overnight bus for the eight-hour drive to CTIO headquarters, at which point he'd meet our shuttle, ride up to Cerro Tololo, and stay up all night observing.

Luckily for him, it hadn't worked out. He got a different flight from Miami, so instead of the bus, he was taking the plane after all. We stopped at the airport to wait for him (this was how Tyson, Kirshner, and John Tonry ended up holding their miniconference on the weather in the parking lot). The plane finally came in, and Raja wasn't on it. Santiago airport was fogged in, and his plane had been diverted to Buenos Aires, where it sat for three hours. He would be getting the day bus after all, and it wouldn't be in until early evening.

So we took off, across farmland where peasants stooped harvesting potatoes, and those whose bags were full waited by the road for a bus to come along. On the far side of an enormous field,

I saw a man guiding a horse-drawn plow. As the land grew more arid, vineyards began to appear; the La Serena area is part of Chile's renowned wine-growing region. Some of the wine is distilled into a ferocious brandy called *pisco,* which is considered part of the cultural experience of Chile; the first question Ed Turner asked me when I got back was whether I had tried it. The answer was no, but most of the astronomers in town had at least one glass the evening before at Mark Phillips's party after Tyson's talk. That Tyson was clear-eyed and cheerful the next morning was undoubtedly a consequence of the fact that he had switched to Coke after the experience.

The shuttle continued up the highway, a two-lane paved road that paralleled a mile-wide, nearly dry riverbed for almost an hour, and then turned off onto a dirt track. We went through a gate (the shuttle driver left several loaves of bread off with the attendant, who lived at the gatehouse) and began the half-hour ascent to Cerro Tololo. The landscape looked remarkably like that around Mount Hopkins, although it was a little less dry. The road was much smoother and better maintained than the road to Mount Hopkins, but Tyson told me that the first two telescopes on the mountain were originally brought up by burro. The road wove in and around hills, nearly always climbing, and the domes on the nearby mountaintop kept coming into view from behind the foreground hills, then disappearing again.

At the top, we were issued flashlights, parkas, and room keys and then, rushing to get there before the cafeteria closed, went on to lunch. The dining room was spectacular. The east wall had floor-to-ceiling windows, and outside the windows there was only a narrow shelf a few feet wide and then a sharp downslope. The combination of the high Andes above and the valleys below made it seem as though we were floating unsupported at seven thousand feet. "There are condors that nest up here," Tyson told me. "The kitchen staff used to feed one of them right out here on the ledge." Before we left the cafeteria, we signed up for night lunch, a bag of sandwiches and cookies the kitchen staff puts together for any astronomer who is working through the night. Then we hiked up to the dome, a longer and more difficult walk than the equivalent trek at Mount Hopkins since the straightest route is up a gravel path.

High overhead, a thick layer of cirrus clouds was making Tyson nervous. "This isn't the kind that blows away when the sun goes down," he said. "Usually, our kind of observing is punishing. We have to be incredibly careful about calibrating the camera, since we're trying to go so faint. So we have to spend the whole after-noon taking flat fields before we spend the whole night observ-ing." A flat field is a snapshot through the telescope of a blank white screen on the inside of the dome; since the CCD is taking a picture of something pure white, any dark spots are by default defects in the camera itself, and the computer can subtract them later from sky images. "We'll often work from right after lunch right through until the next morning," said Tyson. This is partly why he had been so anxious about when Raja would show up: he needed the help. Tonight, though, he would probably not be doing much astronomy, so he would take the flat fields later. He talked with the telescope operator about what they'd be doing, asked some questions, and then went to get some sleep.

Over dinner, Tyson insisted that I try the dessert, a strange-looking fruit called cherimoya. He loved it so much that he had once brought seeds back to New Jersey and grown a little cheri-moya tree at his house. It survived for several years but never bore fruit. I tried it and decided to stick with cookies. Just as I had done with Huchra, I asked Tyson what he would do if a supernova went off during the run. "I know exactly what I'd do," he said, "since it already happened to me once." He and his wife were on the four-meter one night in February 1987, when the light from the exploding star in the Large Magellanic Cloud reached Earth, the first nearby supernova in centuries.

"As it happened," he said, "no one on the mountain, especially us, knew anything about how to observe it, so we ran to the library and frantically read up on the subject. Pat and I were all set up to do spectroscopy of twenty-fourth magnitude galaxies and suddenly had to take data on a second magnitude object. I put as many filters as I could over the detector, but that wasn't good enough. So I defocused the telescope. The light was still so strong that we could take a spectrum every few seconds. We could see the shock wave expanding at sixteen thousand kilometers a second. It was really spectacular. Eventually, you could see the shape of it changing, and you could see the elements come and go—oxygen, calcium, helium, sulfur. We made a motion picture of the spectral lines changing.

"It's really good for me to collaborate with Pat from time to time. She tells me about something exciting going on in her field, and I say 'Oh?' It makes you realize—you think you're really understanding the universe, but you're really only understanding one small aspect of it. And sometimes you get new insights on your own work. Her knowledge about cool, low-mass stars combined with my deep imaging is how we could set limits on how much of the dark matter can be accounted for by dim stars."

Back up in the dome itself, the telescope was ready to go; all we needed was darkness and the break that had appeared in the clouds to widen. We went into the dome. The mirror, thirteen feet across, had a big, dark blotch on one side of its silvered surface. "Looks like somebody spilled coffee on it," said Tyson. It wouldn't make much difference to the observations: the blotch would dim the starlight, but it was small in relation to the total area of the mirror.

The operator on duty that evening spoke English, but with a heavy Spanish accent. Tyson told me that academic jobs in Chile are so hard to come by and so low paying that several of the controllers—not this one—are former physics professors who want to move a rung up the socioeconomic scale. Tyson hunted around the control room until he found a sheet of clear plastic and a roll of masking tape. He taped the plastic over the video screen that showed the view through the telescope.

This was an essential part of his technique for taking ultradeep images. Unlike the relative avalanche of photons that reaches a telescope from nearby galaxies, the faintest blue galaxies only supply about one photon a minute. This would not be a problem, since the CCD will capture and store each of them. But despite how it looks to the human eye, the sky itself, even between the stars where it looks utterly black, is only relatively dark. At its darkest, the sky is still about six hundred times brighter than a thirtieth-magnitude galaxy, thanks to sunlight and starlight bouncing off atmospheric and interstellar dust. For every blue-galaxy photon arriving from deep space, there are six hundred photons of junk that are faithfully recorded on the CCD as well.

To get around this seemingly intractable problem, Tyson invented an entirely new technique; that in fact is how he discovered the faint blue galaxies in the first place. In essence, he jiggles the telescope. He takes one exposure and then moves the telescope just a fraction before taking another. "Imagine you're looking through a dirty window," he explained, "except that in this case, the dirt is light from the night sky. When you move the telescope, the garbage all stays in the same place, since it's random, like static. But the real stars and galaxies move. That eliminates them completely from each individual exposure—they're not in the same place long enough to build up an image. What you are building up is an image of the dirt. Finally, when you've got a good image of the dirt, you take it and eliminate it electronically from each image. Then you reregister the exposures, and when you recombine them the images of faint objects stand out—they add up, while the dirt is mostly gone. To make this work, we do thirty or forty exposures of each field, of about five hundred seconds each.

"I was forced to invent the technique because of the horrible quality of the early CCDs. The general idea was floating around—radio astronomers do what they call 'chopping,' which is subtracting a flat field from an image. Pat Seitzer and I perfected this in 1982 or '83, and now it's starting to be used by others." The reason for the plastic he had taped over the screen was so he could use a grease pencil to mark the position of a guide star during each image. "The computer remembers where the telescope was pointed, of course," he said, "but this helps me to remember, too. It's lucky I found the tape. You know, no great discovery in astronomy was ever made without masking tape."

As the darkness came on, and the clouds stayed reasonably thin, the first order of business was making sure the telescope was in focus. The process was more painstaking here than the manual operation at the MMT. "You can get an approximate focus by hand," he said, "which is okay for spectroscopy of nearby galaxies. But for imaging very faint objects, it isn't really quantitative enough." So he and the telescope operator took a series of exposures of a star with the telescope set at several points in and around what they thought the ideal focus might be. They snapped an image of each with the CCD, and then had the computer display a graph of how concentrated the starlight was. The sharper the slope of the graph, the more light there was in the very center of the image and thus the better the focus. They had to do the operation for each of the filters Tyson would use to make his images: one each that allowed blue-ultraviolet, red, and near-infrared light through. "The precise colors we use are chosen both to bring out the foreground and background galaxies and to filter out the glow from the atmosphere as much as we can," he said.

"Okay, let's go," he said. "Time to go to the first object." The telescope slewed into position (this building stayed still) and Tyson and the operator squinted at the screen and then at their finding chart, trying to figure out what they were looking at. "Okay," said Tyson, "try moving west by two arcminutes." The field of view slid over and a bright star came into sight. The telescope stopped, and the image of the star wobbled back and forth a few times as the telescope settled down. Then Tyson typed the observing command into his terminal, and the integration began. While the photons silently accumulated, he marked the star's position on the plastic overlay. Several minutes later it was over, and it was time to move just a bit. Tyson reached for the fine-movement controller, a device that looked like a control for a video game. Moving the telescope over very short distances is considered a safe enough procedure that even an astronomer can handle it. He squinted at the screen, goosed the control, and moved the star over an inch or so on the screen. He marked its new position and began the second exposure. "This is really exciting, isn't it?" he asked as we waited. "It's like waiting for a train."

Suddenly, he remembered something. "We never picked up our dinners. I'll have to go back down. I find that if I screw my head on one extra turn, it sometimes helps. But you want to be careful with that. You can strip the threads." The kitchen was too far away for a round-trip on foot between repositioning, so we drove down on one of CTIO's Volkswagen bugs. The clear patch of sky was still there, but it was growing no bigger. I wanted to see the Southern Cross, one of whose stars is Alpha Centauri, but it was behind a cloud bank. I could see the Milky Way, though, and even through the thin overcast it was much thicker and brighter than it looked from Arizona: I was looking toward the center, where both the star clouds are thickest and where, but for the dust lanes that get in the way, you would be able to see the brilliant heart of the galaxy. (One night at nearby Las Campanas, Sandra Faber told me during a break at a conference, the sky was so clear and the Milky Way so bright that she briefly transcended the two-dimensional illusion. "I could actually see the sky in depth," she said. "I felt as though I were suspended in space, and I really felt the enormous distances.") Tyson is no John Huchra when it comes to naked-eye objects, but he did find the Large Magellanic Cloud for me, just 150,000 light-years away compared with Andromeda's 2 million.

Back in the dome after picking up the dinners, he was ready to go to the second cluster. "I have a feeling this one isn't very compact," he said. "In fact, I think it's possibly a superposition on the sky of two different clusters. Sometimes when you go trolling for albacore, you get a few sharks." Periodically he would go outside and check the sky; it didn't get much worse, but it didn't get better either. Tyson would hang in for the rest of the night, but I had to get a plane back to Santiago in the morning. I found out

M31, the Andromeda Galaxy   COURTESY NOAO

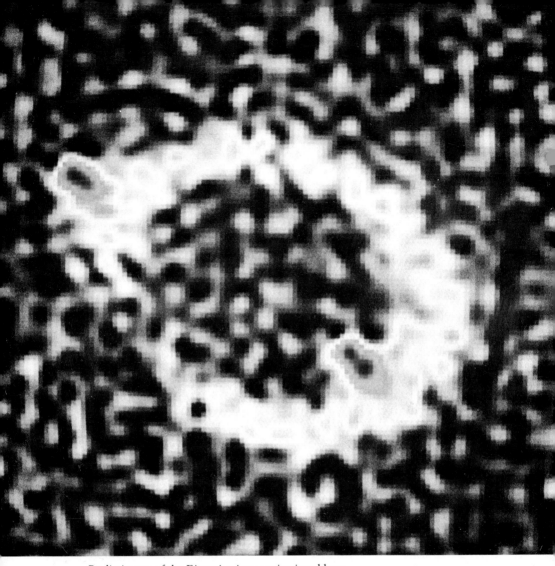

Radio image of the Einstein ring gravitational lens
COURTESY JACQUELINE HEWITT / NRAO

Hotspots in the light at the edge of the universe, imaged in microwaves by the
Cosmic Background Explorer satellite

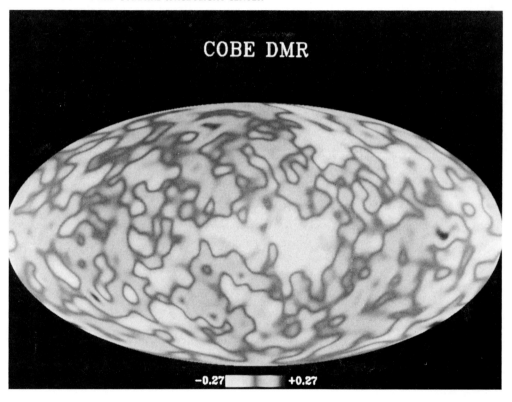

David Wilkinson and the Princeton microwave-background detector, Green Bank, West Virginia, 1991   PHOTO: MICHAEL D. LEMONICK

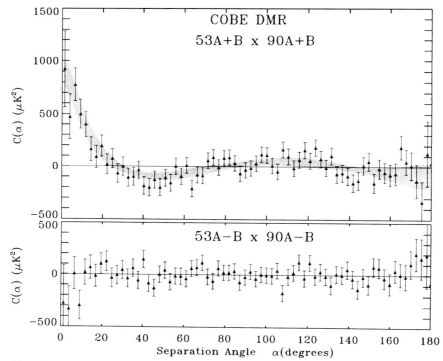

Chart showing how density fluctuations in the early universe vary with angular separation. The gray band is an ideal Harrison-Zel'dovich-Peebles spectrum. The COBE data points are marked by triangles, with vertical lines representing the margins of error.   DIAGRAM COURTESY OF GEORGE SMOOT, REPRODUCED WITH PERMISSION FROM *THE ASTROPHYSICAL JOURNAL*, SEPT. 1, 1992

Radio telescopes and the international no-sparkplug symbol at the National Radio Astronomy Observatory, Green Bank   PHOTO: MICHAEL D. LEMONICK

MASS-DENSITY IN SUPERGALACTIC PLANE Wv, Rw=R5

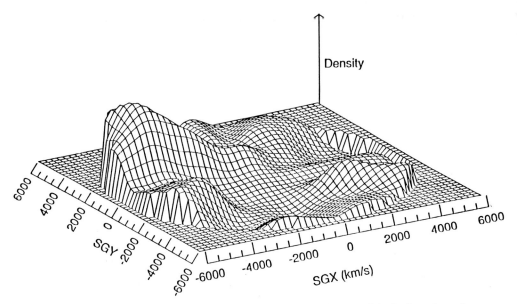

Plot of mass density in the supergalactic plane; the highest peak is the location of the Great Attractor.   COURTESY OF SANDRA FABER

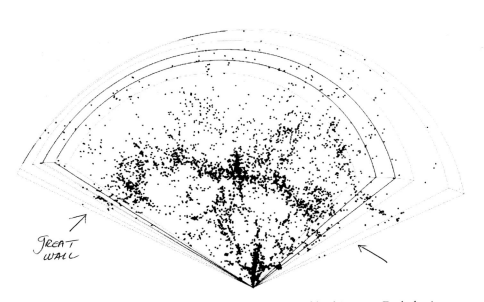

A slice of the universe, centered on the Milky Way and looking out. Each dot is a galaxy. The empty spaces are voids; the Great Wall is the broad band of galaxies that spreads across the entire slice.

COURTESY HARVARD-SMITHSONIAN CENTER FOR ASTROPHYSICS

The 60-inch telescope on Mount Hopkins, in Arizona, where most of the slice of the universe was charted   COURTESY SMITHSONIAN ASTROPHYSICAL OBSERVATORY

Gravitational lens 1208
+101, discovered by
John Bahcall *et al.* in the
Hubble space telescope
snapshot survey
COURTESY JOHN BAHCALL/NASA

The Multiple Mirror Telescope, Mount Hopkins, Arizona
COURTESY SMITHSONIAN ASTROPHYSICAL OBSERVATORY

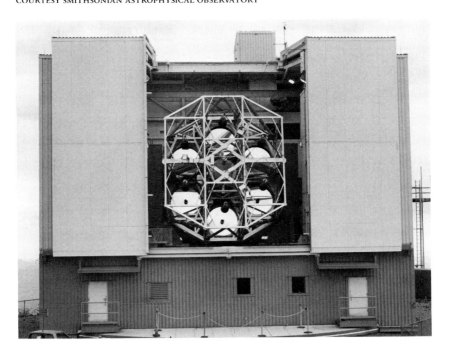

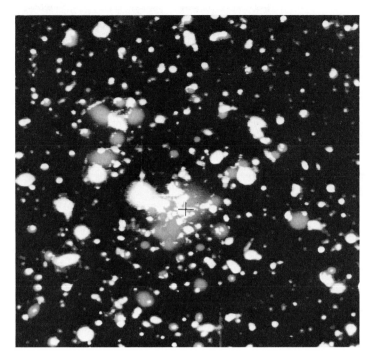

A cluster of foreground galaxies lensing background galaxies
(above) and the distribution of dark matter in the cluster
deduced from the lensing pattern
COURTESY J. ANTHONY TYSON / AT&T BELL LABORATORIES

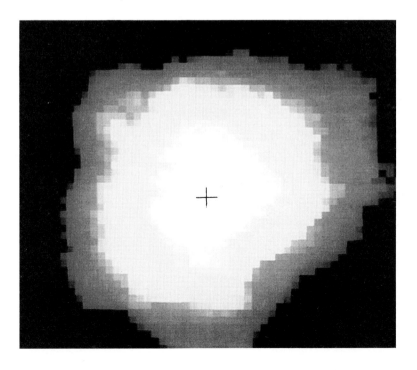

David Wilkinson and Lyman Page making bubbles for William Page
PHOTO: MICHAEL D. LEMONICK

Opposite, top:
Bookshelf in Robert Dicke's office, showing Dicke's dissertation and those of his
students, including Jim Peebles   PHOTO: EILEEN HOHMUTH-LEMONICK

Opposite, bottom:
From left: Freeman Dyson, Alan Guth, Timothy Ferris, J. Richard Gott at the
Princeton COBE conference, June 1992   PHOTO: MICHAEL D. LEMONICK

R.H. DICKE 1941
G. NEWELL 1953
G.S. NEWELL 1953
W.B. HAWKINS 1954
J.P. WITTKE 1955
G. SHERMAN 1955
R.H. ROMER 1955
L.D. WHITE 1956
P.L. BENDER 1956
R.B. GRIFFITHS 1957
E.D. LAMBE 1959
C.T. MURPHY 1959
E.D. LAMBE 1959
P.J.E. PEEBLES 1961
C.H. BRANS 1961
W.F. HOFFMANN 1962
C.O. ALLEY 1962
J.W. BRAULT 1962
J.Q. STONER 1963
J.E. FALLER 1963
W. WILDRETH 1964
L.M. JORDAN 1964
W.J. MORGAN 1964

Supernova 1987A *(at arrow)*   COURTESY NOAO

Opposite, top:
Kitt Peak National Observatory, near Tucson, Arizona   COURTESY NOAO

Opposite, bottom:
Telescope domes at Cerro Tololo Inter-American Observatory, Chile.
The four-meter is second from left.   COURTESY NOAO

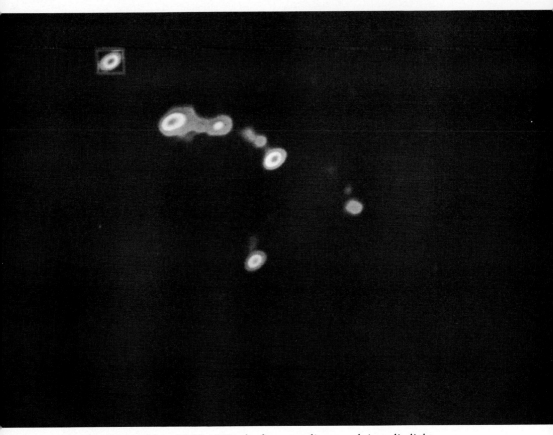

Gravitational lens 0957 +561, the first ever discovered, in radio light

The Very Large Array radio telescope near Socorro, New Mexico, in its most compact configuration   COURTESY NRAO

Karl Jansky's antenna, the first radio telescope   COURTESY NRAO

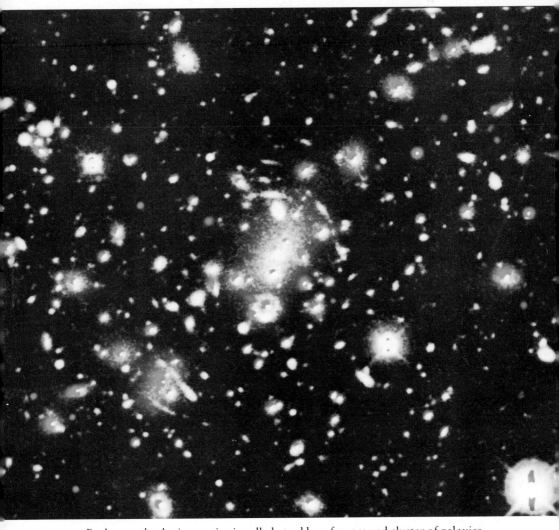

Background galaxies gravitationally lensed by a foreground cluster of galaxies. Lensing stretches the background galaxies into arcs, or fragments of a circle. Several can be seen close to, and especially above, the cluster's bright core.

later that the rest of the run was even worse than that first night, that Raja made it up to the mountain on the second night and that Tyson, for the first time ever, hadn't gotten sick.

Tyson roused himself from bed to see me off at noon the next day and showed me one more sight before I left. He pointed to another mountain, perhaps five miles away. "That's Cerro Pachon," he said. "It's about ten thousand feet high, and a site survey has shown that the seeing is even better there than it is here. That's the site the National Science Foundation picked for the new eight-meter telescope they were going to build. They had one planned for Mauna Kea, in the Northern Hemisphere, and one for here. Now they say they're building only the one in the north, unless we can come up with a foreign investor to share the cost of the southern one. Meanwhile, the Europeans are going ahead with plans to put four eight-meter telescopes in at La Silla. It's really a terrible shame."

CHAPTER SIX

HOW OLD IS THE UNIVERSE?

I f Tony Tyson ever got his wish and managed to convince some improbably generous benefactor to fund the construction of an optical telescope with a mirror twenty-five meters in diameter, it would be an impressive device, with a light-gathering area more than six times as great as that of the Keck telescope, recently completed atop Mauna Kea. But for sheer size, it wouldn't compare with the telescopes radio astronomers use to study the low-frequency radiation emitted, along with visible and other types of light, by stars, galaxies, and quasars.

Where light waves are measured in hundred thousandths of an inch, the smallest distance between the peaks and troughs of radio waves is on the order of an inch, and the concave dishes that catch and concentrate these waves must be correspondingly bigger. The 140-foot radio dish that towered over Suzanne Staggs's microwave antenna at Green Bank, West Virginia, is almost twice the size of Tyson's dream telescope, but it is far from being the world's largest. A mile or so farther into the woods at the Green Bank site there used to be a radio dish 300 feet across. It collapsed one night in 1989, leaving intact only the brick control building that stood under it. The operator on duty didn't realize what had made such a loud noise until he walked outside and saw the telescope lying in pieces all around him. Thanks to Robert Byrd, the bluegrass-fiddling, pork-barrel politicking senator from West Virginia, the 300-foot is being rebuilt, a bit larger than before. There is a 250-foot radio telescope at Jodrell Bank, in England. There is a 300-foot-plus radio telescope near Bonn, Germany. And in the Montanas Guarionex range of northwestern Puerto Rico is the Arecibo radio telescope. Unlike the others, it is far too big—1,000 feet across—to be mounted on a platform and swung to point at different sections of the sky. It rests on the ground, fitted into the contours of a bowl-shaped valley and pointing perpetually straight up, while a 525-ton platform carrying detectors to receive the focused radio waves sits above it, suspended from three massive cables.

Arecibo is the biggest single radio dish in the world, and it is unequaled in its sensitivity; the observatory is a mecca for radio astronomers who want to study the faintest wisps of radio energy wafting in from the universe. Martha Haynes and Ricardo Giovanelli spend so much time working at Arecibo that they have bought a house near the telescope. When John Huchra wanted to collaborate on a project to measure the distances to galaxies by looking at the rotation rates of their radio wave–emitting gas clouds, he signed up with someone who had observing time at Arecibo.

Yet sensitivity to extremely weak electromagnetic radiation is only one measure of a telescope's power. Another is resolution: how precisely can a telescope distinguish small structural details in the objects it is aimed at? The human eye has its own resolution limits. The middle star of the handle of the Big Dipper is really a quadruple star—two pairs of double stars orbiting around a common point. The reason our eyes see just one star where there are really four is that they're too close together. They can't be resolved.

The human eye's powers of resolution are limited by its size and the materials it's made from. Big optical telescopes, which can be made large and almost perfectly smooth, are limited by the shimmering of the Earth's atmosphere. The atmosphere is peppered with pockets of air that are either a little cooler and denser or a little warmer and more tenuous than the average. Each pocket acts as a kind of lens, distorting the starlight passing through. The little pockets constantly shift and jump, making stars twinkle and astronomers curse. As a result of the former, even in high, dry sites like Cerro Tololo, where atmospheric turbulence is at a minimum, objects any closer together than about one arcsecond on the sky—a half-arcsecond, at the absolute best—appear to merge together. That was what made the Hubble space telescope worth billions of dollars and years of development. Even though it has a modest 2.4-meter mirror, the Hubble is flying above the atmosphere and is thus immune to its distortions. If its mirror had been ground to the right shape, the Hubble would have been able to separate objects only a tenth of an arcsecond apart when looking in visible light, and half that in ultraviolet. Despite what NASA's overzealous public-relations machinery claimed, the Hubble's great strength was not that it could see seven times farther than earthbound telescopes but that it could see more sharply.

Radio waves, on the other hand, don't care much about the roiling atmosphere. They pass right through without distortion. The sharpness of a radio image is limited only by the size of the telescope—the bigger, the better. Jodrell Bank does better than the 140-foot dish at Green Bank, and Arecibo does better still. But if Arecibo is the mecca of sensitivity, the mecca of resolution is in central New Mexico, in a vast desert valley called the Plains of San Agustin. Instead of trying to build a single radio dish bigger than the one in Puerto Rico, scientists at the National Radio Astronomy Observatory (NRAO), have installed twenty-seven radio telescopes, each one eighty-two feet across, that work together to simulate a single telescope. Collectively the dishes are known as the very large array (VLA), and the idea is analogous to (but in some ways totally different from) what the Smithsonian Observatory did in building the multiple mirror telescope on Mount Hopkins.

Astronomers usually approach the VLA from the east, since the road from the nearest airport, in Albuquerque, and from the observatory headquarters, in the town of Socorro, comes in from that direction. I made the approach late one afternoon in midsummer in the company of an astronomer named Jacqueline Hewitt and her graduate student, Grace Chen, and as we drove into the sunset, I kept craning my neck to get a first glimpse of the site. The valley is desert, but a slightly lusher desert than the country toward the east where we had come from. Sparse, scrubby bushes had given way to dryland grasses and small trees, and the trees kept getting in the way of the view. Noticing my impatience, Hewitt said, "You should be able to see it as we come over that next rise. But you might be disappointed. The telescope is in the A array right now. That means it's at its maximum size." The twenty-seven dishes of the VLA are divided into three sets of nine, each set forming one arm of a Y. The arms are delineated by railroad tracks that allow the dishes to be moved closer together or farther apart, depending on what kind of observations are going on. "When it's in the A array," said Hewitt, "each arm is twenty kilometers long. It's kind of hard to see much of it."

We topped the rise and there, spread out before and below us, were the valley and the mountains that surround it. The road stretched out in front of us, absolutely straight as it crossed the valley, finally disappearing in haze at the base of a mountain range on the other side. It took the shortest path across the valley. "The town of Datil is right there, where the road meets the mountains," said Hewitt. "From the VLA control room down there, at night, you can see the headlights of cars on this highway. You'll see the lights over on one side of the valley, and then you look a half hour later and you see the same car, finally reaching the other side."

I scanned the valley, looking for the world's biggest radio telescope, and saw nothing for a few moments. Then I noticed a tiny

Jacqueline Hewitt   PHOTO: EILEEN HOHMUTH-LEMONICK

white dot off to the left, like a single mushroom in the middle of a football field. It was backlit by the setting sun so that it was mostly in shadow. Once I realized what an eighty-two-foot radio dish looked like from this distance, I suddenly began to pick out more of them. I got all the way to four. The other twenty-three were effectively invisible, and I could not imagine where they could be hiding, even in an area this large. As we descended to the valley floor, other dishes finally began to appear, but the farthest were lost in the failing light.

Finally, we reached a small sign pointing to a side road that goes to the VLA. "Before they put up the sign," said Hewitt, "it used to be easy to miss this road. One night a few years ago, Bernie Burke [the same man who first put the cosmic microwave background theorists at Princeton in touch with Arno Penzias and Bob Wilson at Bell Labs in 1965] was driving a group of us back from dinner in Datil. He got so involved in the story he was telling that we shot right past the road. He turned the car around, but by the time we got back to the turnoff he was into another story, and we shot by again. It took about five tries to get it." We had no such mishap, and we successfully avoided the potential for another: a cow was grazing right next to the side road. "Sometimes they just lunge out in front of you," said Hewitt. This one didn't. The cows are part of the scene at VLA. The land is privately owned and ranched, and the NRAO has made arrangements to use it. "The cows sometimes walk over and scratch their backs on the antenna supports," she said.

Hewitt knows the VLA intimately, in all its configurations. She is there frequently, working on radio observations of all kinds, but she is particularly interested in the same phenomenon that has grabbed Tony Tyson's attention. She is a world-class radio observer of gravitational lenses. Different wavelengths of electromagnetic radiation behave differently in many ways—some penetrate dust clouds, others don't; some are absorbed by water vapor, others aren't. In their response to gravity, though, they are identical. Gravity is color-blind. Unlike, say, a prism, which bends different colored light through different angles, the gravity of a massive object will bend red, blue, and yellow light by precisely the same amount. It will do the same to radio waves, microwaves, and X-rays, which are essentially colors of light the human eye can't see.

Tyson's faint blue galaxies would be visible in a radio telescope if they emitted much in the way of radio waves. Most of them don't. Astronomical radio waves are usually generated by electrons getting thrown around at high speed, and that kind of violence is unusual in ordinary galaxies, but not in the objects, particularly quasars, which Hewitt studies. Radio observations of gravitational lenses are useful for two reasons. First of all, because of the indifference of gravity to the kind of light it's bending, a radio observation that shows two or three images of a single quasar can then be confirmed by an optical observation that shows the same thing. If it's there in the optical but not in the radio, it may be interesting, but it's not a lens (unless the quasar happens to be radio quiet, in which case the lack of a radio signal is useless). Second, a big radio telescope's ability to resolve fine details in the structure of an astronomical object or of an astronomical illusion lets astronomers understand what is going on in the lens system better than they could with optical images.

Hewitt latched on to lensing early in her career, while she was still a graduate student at MIT, and now that she's on the faculty there she is considered one of a handful of experts on the subject. The American Astronomical Society paid her the professional compliment of asking her to deliver an address on lensing at its semiannual meeting in Seattle in June of 1991. (She was nervous and unsure both before and after the talk, but despite a microphone that refused to work and a large, packed auditorium, she impressed the assembled astronomers.) Like Tyson, Hewitt had gone beyond simply finding examples of gravitational lensing in the sky, and was now trying to use them as a tool to answer one of astronomy's long-standing mysteries. Tyson's mystery—the question of where and what the dark matter is—dates back to the 1930s, but Hewitt and the handful of astronomers she collaborates with were interested in a problem that is probably as old as human history itself: how big is the visible universe, and how old?

These sound like two separate questions, but essentially they're not. Astronomers already know how fast the galaxies are flying apart: the redshifts they measure are precise indicators of recessional speeds. If they knew the galaxies' distances equally well, it would only take simple algebra to figure out how long it took, from the time when the galaxies were all squashed together in the Big Bang, for them to get that far apart. Figure out the size, and you know the age. In fact, it's not quite that simple. What you know at this point is something called the Hubble time, the time back to the beginning of the universe *if* the universe had no mass. But of course the galaxies weigh something, as does the dark matter. All that mass exerts gravity, and gravity slows the expansion. To get the real age of the universe, you have to know how much mass there is and correct for the slowdown. If the cold dark matter crowd is right, and there is just enough matter to stop but not reverse the universe's expansion, then the real age of the universe is two thirds of the Hubble time.

Knowing the size and age of the universe could either help resolve the cosmological crisis or make it worse, depending on what the answer is. The current range of sizes astronomers have placed on the visible cosmos is ten to twenty billion light-years in radius. The "right" answer (that is, the desirable answer) is that the edge of the universe is twenty billion light-years away. That makes the Hubble time twenty billion years, and if cold dark matter (CDM) is right and omega is one, the real age thirteen billion years. The answer is desirable because an older, more slowly expanding universe is more likely to form large structures—such as Great Walls and Great Attractors. It's also desirable because there are individual stars within the universe, located in the globular clusters that orbit spiral galaxies like Andromeda and the Milky Way, which are, without much doubt, at least twelve or thirteen billion years old. If the answer is "wrong," the Hubble time is only ten billion years, CDM says the age is only seven billion years or so, and the universe is therefore considerably younger than some of the objects in it. The majority of astronomers who specialize in distance measurements are getting the wrong answer.

Thanks to Hubble's law, it should be easy in principle to answer the question of how big the visible universe is. If galaxy A is moving twice as fast as galaxy B, it's twice as far away. You can make a graph with speed along one axis and distance along the other, and just about every galaxy should fall on a single, diagonal line on the graph, a line that moves from lower left to upper right. The line won't be accurate for local galaxies, because Hubble expansion velocities are small, and the "peculiar motions" induced by the gravity of nearby galaxies are comparatively large (as in the case of Andromeda, for example, whose motion away from the Milky Way in the expansion of the universe has been completely overcome, and which is actually approaching rather than receding).

But out where the speed of the Hubble flow is great, peculiar motions hardly count; there, Hubble's law should predominate; you should be able to measure the redshift of a faraway object, a quasar, or even the microwave background, convert that measurement into a speed, and convert the speed into a distance. The trouble is that no one knows precisely what recessional speed corresponds to what distance. Nearby, where it's easier to measure distance, redshifts are misleading. Far away, where redshifts are useful, it's hard to measure distances.

It wouldn't be hard if the cosmos had supplied astronomers with a few good hundred-watt light bulbs. The thing about a hundred-watt bulb is, you know how bright it is. It might not look very bright if it's far away, but that doesn't matter. As long as you know how inherently bright an object actually is, the laws of physics let you use its apparent brightness to calculate its distance. A good, standard hundred-watt cosmological bulb (or, as astronomers prefer to say more quaintly, a "standard candle") would answer the question right away. If all spiral galaxies were equally bright, of if all ellipticals were, or if all supernovas were—anything that can be seen well into the depths of the universe—and if you knew what their inherent wattage was, you could look at one halfway across the universe, calculate how far away it was, and then use that number to calibrate Hubble's law, the redshift-distance relationship that is a consequence of the expanding universe.

It's hard to find a standard candle, though, or at least one that can be seen far enough away to make a difference. Supernovas are not all equally bright. Galaxies might be, if astronomers were always careful to use exactly the same subtype all the time. But if you go out very far into the Hubble flow, you're looking at galaxies when they were younger, and galaxies evolve. There are good arguments that say they get dimmer with time, and good arguments that say they get brighter, and no one knows which effect wins.

It is the lack of a standard candle, or of some other way to measure directly the distance to a significantly redshifted galaxy, that has had astronomers arguing back and forth over the past three decades about what the calibration number for Hubble's law, known as the Hubble constant, really is. The Hubble constant is essentially a single specification for a contour map of the universe. With every megaparsec farther from Earth you go, every 3.26 million light-years, the galaxies are flying away faster by some constant number of kilometers per second. Those who argue for a high Hubble constant think the galaxies are separating one hundred kilometers per second faster with every megaparsec. That is the same as saying the universe is young; moving faster, it took less time to reach its present size. Those who say the Hubble constant is fifty or so are presenting evidence for an older universe.

In the absence of a hundred-watt bulb, astronomers have relied since the days of Edwin Hubble on a stepwise approach to measuring the universe. They calculate the distance to nearby objects, then use those distances to gauge the distance to further objects, and so on, out into the cosmos. Unfortunately, each of the steps on this distance ladder is only approximately accurate. The uncertainty is minor at each step, but the uncertainties add up over many steps. That is why the final estimate—the Hubble constant, and thus the size of the visible universe itself—can vary so widely.

The most accurate step on the measurement ladder is the first one: the distance to stars within one hundred light-years, by means of trigonometric parallax. The easiest way to understand parallax is to hold up a finger in front of your nose, and close first one eye, then the other. The finger seems to jump across your field of view. It works the same way with stars. Look at a nearby star in, say, January and note its position against the general background of distant stars. Look at it again in July; the distant stars won't appear to move, but seen through the eyepiece of a telescope at least the nearby star will. It's easy to calculate how far away it is by how much it moves.

Beyond a hundred light-years, the technique doesn't work. The shift is still there, but too small to be detected through even the most powerful telescope, so astronomers go on to the next step. The color of a star is a reasonably accurate measure of both its size and its age, and using parallax-based distances to nearby stars of various colors astronomers can calculate their inherent brightnesses. The stars become standard light bulbs of varying, but known, wattages, allowing astronomers to gauge the distance to their similarly colored cousins far into the Milky Way. Because the technique is only approximate, astronomers generally use clusters of stars rather than individuals. All the stars in a cluster are obviously equally far from the Earth, so the inevitable variations in distance measurements can safely be averaged to refine the accuracy still further.

The next step on the distance ladder is the Cepheid variables, stars whose brightness rises and falls with a regular rhythm of several days to a few weeks (in fact, this represents a stage late in the life of most stars, not a particular type of star; the name comes from the fact that the first one was discovered in the constellation Cepheus). Around the turn of the century, the Harvard astronomer Henrietta Swann Leavitt discovered that there is a direct relationship between the length of a Cepheid's period and its intrinsic brightness. You can tell by how long it takes a Cepheid to flicker how inherently bright it is, and thus how far away it is. It was Cepheids that let Edwin Hubble prove that the fuzzy nebulas floating in his telescope's field of view, the Andromeda nebula in particular, were far beyond the Milky Way.

The next step takes distance measurements all the way out into the Hubble flow, where redshift finally becomes an accurate indicator of distance. Hubble took this step by guessing that Andromeda was a typical spiral and by comparing its brightness with that of more distant spirals. It was a reasonable guess, but Hubble had measured the Cepheid distance to Andromeda incorrectly; the value he got for the Hubble constant was five hundred kilometers per second per megaparsec, and the age of the universe about two billion years.

Still, his basic technique was sound. By the 1960s, astronomers had sharpened their measurements of the ladder's steps. By the early 1980s, they had abandoned the direct brightness comparisons between nearby and faraway galaxies as well. It turns out that the inherent brightness of a given spiral galaxy is closely related to its rotational speed, and that brightness of an elliptical, which doesn't rotate, is related to the average speed of its stars. The former is known as the Tully-Fisher relation and the latter as the Faber-Jackson.

It was the Faber-Jackson relation that led to the discovery of another chink in the armor of the standard cold dark matter theory. Having helped come up with the relation, Sandra Faber is the woman who convinced George Blumenthal to take dark matter seriously. With Blumenthal, Primack, and Martin Rees she wrote some of the earliest papers on the topic, and went out to test her relation on galaxies in the local hundred or so megaparsecs of the universe. On various observing runs at different telescopes over several years, she and six collaborators, the Seven Samurai, painstakingly measured the distances to several hundred galaxies. Once you have the distance and the redshift of a galaxy, you can plot it on the line that marks the Hubble constant on a distance-velocity chart. You can if it falls on the line, that is, but these galaxies didn't.

Faber herself explained to me what happened next. I was lucky enough to find her in her office and not out observing when I went out to Santa Cruz to see Blumenthal for the second time, and even luckier that she had a few minutes between writing up grant proposals and leafing through the latest space telescope photos (she's got several observing projects going on the Hubble). She's the kind of relaxed and friendly person who doesn't mind visitors poking their heads into her office. Her official publicity photo shows her in a business suit, with an enormous smile on her face.

In person, the smile was exactly the same, but she wore jeans and a blue sweatshirt from Swarthmore, her undergraduate school. Faber is on the shorter side of average and wears her blond hair short.

"We were certainly not the first ones to measure peculiar motions of galaxies," she said, "but we were the first to map them in terms of the cosmic rest frame of the microwave background." That is, instead of treating the local group of galaxies as being at rest and the rest of the galaxies as flying away—not an unreasonable approach—the group decided to treat the edge of the universe as fixed and see how everything else was moving in relation to it. It shouldn't make a bit of difference, unless by some chance the local group was not at rest. "Originally," said Faber, "it wasn't

 Sandra Faber    PHOTO: DON FUKUDA

obvious to me that it mattered what rest frame you used, but I remember a moment of epiphany that came when I was talking to Donald Lynden-Bell [another Samurai]. We were walking to lunch one day, and he said, 'I really think we should use the CMB as the rest frame.' It suddenly hit me that if we did, this would make a huge chunk of space that otherwise looked stationary begin to move at about six hundred kilometers per second. Here's a map of what we found."

The sheet of paper she showed me had dozens of small circles on it representing galaxies in the local area, with lines pointing out from them. The directions in which the lines pointed showed which way the galaxies are moving. The lengths stood for speed. Just about all the lines, including the one attached to the Milky Way, pointed in the direction of the constellations Hydra and Centaurus, and beyond: a concentration of galaxies known as the Hydra-Centaurus supercluster, once thought to be pulling on the Milky Way, was moving off in the same direction. If it, too, was moving, there had to be something bigger and farther away drawing it and us. "Based on the data, we hypothesized a group of galaxies out beyond Hydra-Centaurus, which Alan Dressler, on the spur of the moment just before a talk, called the Great Attractor.

"You can't really study these things," she said, "without some feeling for the local topography. It's actually been the joy of my life putting together a mental map of the local universe, covering perhaps a hundred million light-years. I just love the idea of it." (Faber, remember, is the one who can directly sense the three-dimensionality of the universe, despite its apparent flatness.) "Okay, the first thing you have to know is that the Milky Way is part of a huge, flat collection of galaxies called the supergalactic plane. We're embedded in it, just as the Sun is embedded in the plane of the Milky Way. Fortunately, it's at right angles to the plane of our own galaxy, or we wouldn't be able to study it at all. Galactic dust would make it invisible."

"Anyway," she said, pulling out another sheet of paper, covered with dots, "here is a redshift map of galaxies in the supergalactic plane, made from the catalog that John Huchra, the Keeper of the Redshifts, maintains. The catalog is three-dimensional, but this sheet represents a cut we've taken that corresponds to the supergalactic plane. One feature, here"—she pointed to a cluster of dots just above the center of the page—"is the Virgo cluster, which is in Ursa Major, and another is the Perseus-Pisces supercluster. And this area over here [she pointed to a region on the upper left where there were fewer galaxies than average] is the Great Attractor. You don't see many galaxies here because there haven't been very many observations. There's a lot more activity now, thanks in large part to the Seven Samurai's work, and I can't wait to see the latest results when they're added to the catalog. This feature up near the top is the Coma supercluster; the Great Wall goes through it. Behind the Great Attractor [for this, she had to go to another map, on a larger scale] is an even larger concentration of mass that we call the Shapley concentration."

Then Faber pulled out another sort of map, one that represents the average mass of a region of space as a flat plane, with excesses of mass shown as hills on the plane. "You can see here," she said, "that the biggest mountain is the Great Attractor; on this map, the Shapley concentration is off the edge. And here's something else that's interesting." There was evidence on the map of a sort of ridgeline that rose from the flat average to meet the Great Attractor. The Milky Way is down in the flats alongside the ridge. She pointed to a point in the ridge itself. The Virgo cluster," she said, "which is usually thought of as a discrete blob of galaxies pulling other galaxies into it, turns out to be instead just part of the ridgeline leading to the Great Attractor.

"So," she continued, "the current status of the Great Attractor region is this. There have been three more surveys, two by us and one in Australia, and ours at least were motivated by wanting to see whether we could see if galaxies on the other side of the Great Attractor were falling into it the way we are. We wanted to see whether there was a symmetric backflow. One problem we ran into was that there are very few galaxies on the other side. There seems in fact to be a noticeable void. So what's emerging as a great complication in all of this is that the structures are not very symmetrical."

There is evidence from the Australian survey, which goes much deeper than Faber's, showing that the Great Attractor itself is being yanked by the even bigger and more distant Shapley concentration. Some astronomers even discount the existence of the Great Attractor at all, attributing all the peculiar motion to the Shapley alone. "As is usually the case in science," said Faber, "I think the truth lies somewhere in between. Yes, there is a flow toward the Shapley concentration, but there is also a Great Attractor." Her best estimate of the sizes of these collections of galaxies is that the Great Attractor is about ten thousand times as massive as the Milky Way, and the Shapley concentration up to ten times bigger than that.

These tremendous concentrations of mass are problematic for standard CDM, just as the Great Wall is, but their existence is based mostly on the peculiar velocities of the galaxies they're presumably attracting. But calling a velocity "peculiar" presumes that you know how far away it really is, and that its velocity doesn't square with its distance. Yet today, for all the refining of Cepheid distances and the development of the Faber-Jackson and Tully-Fisher relations, the Hubble constant is still uncertain. Each step on the distance ladder is more accurate than it once was, but the small uncertainties at every level add up, and the final number still hovers at somewhere between fifty and one hundred.

Astronomers have always wished they could junk the ladder altogether and find a way to gauge the distances to faraway galaxies directly. Although uncertainty is inevitable even in that kind of measurement, at least there wouldn't be multiple errors piled on top of each other. Then, in 1979, Walsh, Carswell, and Weymann found 0957 + 561. The thing about quasars is that some of them flicker. They'll gradually rise in brightness over weeks or months, then slowly dim again. In a gravitational lens system, though, where there are two or more images of a single quasar, the images won't flicker at the same time. Inevitably, the nearby galaxy that does the lensing is positioned, not precisely in front of the quasar but a tiny bit off to one side. The beam of light that travels around one side of the galaxy, to produce one image, has to go slightly farther to reach earthly telescopes than the beam that produces the other one. If the quasar flickers, the pulse of light will arrive somewhat later in one image than in the other. And because the pulse travels at precisely the speed of light, the difference in arrival times—the time delay—is an accurate measure of the difference in how far each beam of light has had to go. The astronomers already know what the angle between the two images is. Armed with that information and the difference in path lengths, they need just one more datum to deduce the distance to the lensing galaxy: its mass distribution, which is what causes the lensing in the first place.

The difference in path lengths is what Jackie Hewitt had come to VLA to measure. She was looking for a time delay in a gravitational lens, which she hoped would lead to a reliable number for the Hubble constant. She had also hoped that Ed Turner, one of her favorite collaborators, would join her on this run. He had wanted to come, too, but had canceled at the last minute: his schedule was much too busy. Like Hewitt, Turner was an early advocate of using gravitational lenses to solve cosmic mysteries. The two have been working together for years, alternating between optical-telescope runs, where Turner is the expert and Hewitt the assistant, and radio runs, where the roles are reversed. "I almost don't remember how we got together," Hewitt told me during the ride from Socorro, "it was so long ago. When I was a young graduate student at MIT under Bernie Burke, he was pushing the idea of all-sky surveys in radio wavelengths. Ed came up to MIT to give a colloquium on lenses, and he realized that Bernie's survey was an ideal way to search for new ones. That was right at the time I was looking for a thesis topic, and Bernie suggested that I could work with Ed. We went on a run at the four-meter telescope on Kitt Peak, and that's when we realized we liked working together."

The two make a high-contrast pair: Hewitt is in her early thirties and athletic-looking, with shoulder-length, unruly blond hair. Her voice is strong, and she speaks quickly; Turner is extremely slow- and soft-spoken. Margaret Geller once told me that Turner doesn't get the recognition he should in the astronomical community because he's too nice, and doesn't fight for it. This may be true, or it may have more to do with the fact that Geller's personal style is direct, almost confrontational; Turner may only be "too nice" in comparison with her. Turner thinks she's wrong, but that may in turn reflect his own low-key personality.

By the time I went on a run with them at Kitt Peak in 1984, Turner and Hewitt had settled into a relationship like that of a long-married couple: lots of pointed banter hiding deep mutual respect and affection. Hewitt told me that once they had gone on twin runs. Turner joined her for a radio run at VLA, and then they were going to Kitt Peak for an optical observing session. They had decided to drive Turner's rented car the five hundred miles from central New Mexico to southern Arizona. "We had a terrible argument about when to leave," she recalled.

This sounded suspicious. The idea that they could have anything resembling a real fight seemed totally out of character for both. Under pressure, Hewitt admitted that she might have exaggerated. "Well," she said, "it was really more like a protracted discussion. Every hour or so one of us would bring up the issue, and we'd discuss the relative merits of our positions in a conversational way (he wanted to get going, and I wanted to get in as much observing time as I could). Finally it was morning, and we still hadn't agreed on what to do, and since I had wanted to stay, that meant I won." Their spouses—Joyce Turner is a teacher: Bob Redwine, a nuclear physicist—feel totally unthreatened by this intimate professional relationship.

Back in Princeton, Turner prepared me for the run by telling me about the state-of-the-art of gravitational lens measurements. "Gauging the distance scale has always been a fundamental goal of gravitational lens work," he said. "In fact, Refsdal first suggested the technique back in the early 1960s, long before anyone had ever found evidence of lensing. In the past year, there's been a real break in the field. For a long time, even after lenses were discovered, the idea of using a time delay to calculate distances has been talked about as an ideal. But two things have now happened. First, Bill Press at Harvard and his collaborators including Jackie have done a convincing analysis of some existing data on 0957 + 5621, the original double quasar found by Walsh, Carswell, and Weymann. They've decided there is a visible-light time delay of about a year and a half." The reason they had to analyze and decide rather than just measure the time delay is that no telescope can sit pointed at even the most interesting object nonstop for months. What the astronomers have to go on is an incomplete collection of measurements that amounts to dots on a graph, and they have to demonstrate convincingly that a particular curve fits onto those particular dots better than any other curve does.

"The second thing," said Turner, "is that Bernie Burke, Jackie, and some other people have independently come up with a time delay in the radio that's about five hundred and forty days, which is comparable to Press. They're all consistent. A few people are still arguing about it, but essentially everyone else agrees.

"So, we have a measured time delay. Beyond that, there have been initial measurements of the velocity dispersion of the lensing galaxy [in this system, the lensing is done by a galaxy within a cluster, and both galaxy and cluster contribute to the effect]. That lets us calculate the mass of the lensing galaxy, and now we're starting to get some space telescope observations as well. All of these have come together, and with a redshift for the lensing galaxy of point three six, that leads to a Hubble constant of forty kilometers per second per megaparsec. That is, relatively speaking, a low number, and it puts the Hubble time at twenty-five billion years. If you take the prevailing view that omega equals one, then the age is two thirds that of the Hubble time, or about seventeen billion years."

Hewitt and those working with her seem to have the right answer for cosmology, the answer that gives the cold dark matter model some breathing room and makes the cosmos at least as old as its oldest stars. But Turner was not finished. "The thing about 0957," he said, "is that it's kind of a messy system—we don't know exactly how the masses of the galaxies in the foreground cluster contribute to the overall lensing, so there is some uncertainty in the forty. Opinions differ, but you can probably get the number up to eighty if you work at it, which would mean a Hubble time of only twelve billion years. Or you can get it down to ten, which means a Hubble time of a hundred billion years. So it isn't as though we have our answer yet. But it is more or less in the range of what other astronomers have gotten with entirely different techniques, so that shows us that we're in the right ballpark. The technique seems to work. And this has generated a kind of gold fever among lens hunters. Two things are now going on. First, there's been a huge surge of work on 0957, with people trying to constrain the mass distribution in the lens further and get a better idea of the geometry of the system. And second, there is a big search on for other, cleaner systems with time delays. Everyone wants to be the first to get one. That's why Jackie is going to the VLA this summer. I have a lot of confidence in this method, but one example isn't enough. I won't be convinced until we get H-nought for two or more systems."

He and Hewitt both agreed there was already a better one, discovered by Hewitt herself in 1986: the Einstein ring. Fifty years after Einstein predicted that a perfect alignment of foreground mass and background object would create a ring-shaped optical illusion, Hewitt found one during a VLA run. "At first," she told me, "I was sure it was some sort of mistake, some glitch in the hardware or software." It wasn't. In pictures, it looks like a nearly perfect circle of light (albeit radio light; the image is what we might see if our eyes were tuned to a much lower frequency). The circle is two arcseconds across, with two bright blobs glowing on opposite sides of the ring. Others have since looked at the ring with optical and infrared telescopes, and while the ring is big enough to be resolved from the ground, the picture is hard to untangle in both wavelengths; the foreground image may be radio quiet, letting the ring shine through, but bright in other types of light, thus contaminating those pictures.

Even so, Hewitt, Turner, and other astronomers are convinced that the two blobs are twin images of a quasar, and that the ring is a distorted image of a jet of electrons spewing at nearly the speed of light from the quasar's core. There is a second image of the jet as well, but it is almost too faint to be seen. The Einstein ring is a much cleaner system than 0957; the lens itself appears to be a single galaxy whose mass distribution is relatively easy to understand, and the fact that the main image is ring-shaped means that it is almost precisely behind the lens, with little ambiguity about its position. If the quasar that gave rise to the Einstein ring flickered, it would be an ideal system for calculating the Hubble constant. But no one had seen it flicker yet and that was why Hewitt had come to the VLA.

Until a few years ago, radio astronomers wanting to use the telescope had to go to the VLA itself, about fifty miles west of the small university town of Socorro. That was where all the controls were, and the computers used for processing the data as well. Recently, though, the NRAO built an operations center on the edge of the campus of New Mexico Tech in Socorro itself, and many visiting astronomers don't make the trek to the VLA anymore. For the first few days of this run, Hewitt followed suit. She and Grace Chen, who was learning to run the giant instrument, stayed in Socorro and crunched numbers. The reason was mainly logistical, Hewitt told me. "Our observing run isn't until Monday night," she said, "and we don't want to set up for it on the weekend, in case we run into problems and need some technical help from the people out here. The weekends are pretty dead at the site. So we're using the time to analyze some of the data we took on earlier runs. We could just as easily do this work back at MIT, but we've got the time to fill anyway."

For the last two days of the trip, they would head down to the array. "One reason," she told me, "is that I want Grace to get some experience at the site. Another is that I want to check out some new hardware they have down there for VLBI observations." (VLBI, or Very Long Baseline Interferometry, is to the VLA what the VLA is to a single dish. Periodically, radio astronomers all over the world tune their dishes in to a single object, simulating for a few hours a single radio dish thousands of miles across. It is much more difficult than VLA work, though. Since the widely scattered telescopes can't be linked by cable, their signals have to be recorded on tape—videotape works best—and combined later at a single site. If the tapes aren't perfectly synchronized, the data are useless.)

Hewitt has a kind of innocent, wholesome look ("Very deceiving, in fact!" Turner scribbled in the margin of a draft of this paragraph), and so does the Taiwanese Chen, who is overwhelmingly polite. I was thus a little taken aback when on the last evening before the run began, they suggested going to a bar in town. "It's a little rough," said Hewitt. "They sometimes have fistfights." I wasn't thrilled with the idea, but I went along. For a while, circling through the dark residential streets of Socorro, I hoped they wouldn't be able to find it, but suddenly we were in the town square. It had clearly been the center of commerce before the strip grew up along Route 50 at the edge of town, with its fast food, 7-Elevens and motels. The square was deserted, but the bar was still there and open for business. I followed the women inside nervously. "Oh, no," said Hewitt, disappointed. "They've cleaned this place up. It looks like a place for yuppies." It looked to me like a place where yuppies, if there are any in central New Mexico, might go to have fistfights. There were no fights that night, but I warily drank my Perrier out of a beer mug.

The next afternoon, we left for the Plains of San Agustin, fifty miles away through western movie country. It was a Sunday, and the VLA cafeteria would be closed that evening, so we stopped for dinner at the only place open in Magdalena, the only town along the way. Magdalena is a main street and not much else. The town is surrounded by desert and mountains, and it took only a little bit of imagination for me to see it with wooden sidewalks and hitching posts. The cafe we stopped in was a converted bank that looked as though it dated from the late 19th century: it was brick, with high, stamped-tin ceilings and worn wooden floors.

As we ate enchiladas from plastic plates, Hewitt explained how the VLA can take pictures with such impressive resolution. The process is completely different from that at the MMT, where the light from six small mirrors adds up to the light-gathering power of one big one. The VLA can work that way, too, but even when the signals are added together, the radio energy captured by twenty-seven, eighty-two-foot dishes falls far short of the single, one-thousand-foot dish at Arecibo.

The VLA's high resolution comes instead from a technique called interferometry. One of the great and confusing achievements of quantum physics, which was developed in the 1920s, is the discovery that light in all its forms has a dual identity. It is at once a stream of individual particles, called photons, and a series of waves. In the ordinary, macroscopic world, you can't have it both ways. Sound comes in waves, not particles; hail comes in particles, not waves. This is not so in the submicroscopic world of the atom. Which aspect of light's dual nature you observe depends on how you go about trying to observe it. A CCD, for example, expects to see light as individual particles while the VLA looks for waves.

A fundamental property of waves is that they interfere with each other. Imagine two sets of ocean waves converging on a beach from different angles. If the two sets of waves are in phase—that is, if the peaks of the waves arrive at the same point at the same time—what you get is a single set of waves twice as big as each of the originals. If the two sets are out of phase—the peaks of one set merge with the troughs of the other—then the waves cancel themselves out, and you get perfectly flat water.

The same thing happens with radio waves, except that instead of being higher or lower, radio waves will be brighter when they're in phase and dimmer when they're out. When the VLA is pointed at a radio source, like the Einstein ring, each dish receives its signal at a slightly different time, because after billions of light-years of speeding through space the signal has to travel a few hundred extra feet to reach some dishes than to reach others. For every pair of dishes (every one of the twenty-seven matched up with every other one) the arriving signals are in phase by varying degrees. Dish A and dish B, for example, might see signals perfectly in phase, while dish A and dish C see them 10 percent out of phase and dishes A and D see them 80 percent out of phase. In each case, the result—the brightness of radio energy coming out of each pair of dishes—will be different, depending on the dishes' positions.

By itself, the measurement from a pair of dishes is useless, a single number that indicates brightness. It could come from a radio source of any shape. Added together, though, in all possible permutations, the possibilities are narrowed exponentially; only an object of a particular shape can produce a certain pattern of interference. They are reduced even further by the fact that Earth is turning. With each passing minute, as the dishes follow the source in unison, the distances from the source to each dish change and so do the interference patterns. The changes over time are duly recorded by VLA's computers, and what emerges after intense computer processing is a detailed image of the unique celestial object responsible for the signals. The image that comes out is equivalent to what a single radio telescope thirty kilometers across could have produced. In trying to measure the size of the universe, Hewitt will be using interferometry to take a detailed snapshot of the Einstein ring, not just once but several times, months apart, to see whether one of the quasar images has brightened appreciably.

If it does, then she'll begin another series, waiting until the second image does the same. Then she'll have her time delay.

Interferometry can be done, in theory, with optical telescopes as well, but it's much harder. The critical thing in making interferometry work is making sure the signals are combined precisely. Otherwise you get interference patterns created not by the differing arrival times of signals but by faulty mixing. Since optical wavelengths are hundreds of thousands of times shorter than radio wavelengths, it's much harder to mesh them accurately. Nevertheless, two ambitious projects are now under way that depend on optical interferometry. The European Southern Observatory in La Silla, Chile, is building the very large telescope (VLT), a set of four, eight-meter mirrors designed to work together and simulate a thirty-two-meter telescope. And on Mauna Kea in Hawaii, construction has been approved for Keck II, a twin of the ten-meter Keck telescope now nearing completion that will be used for interferometry as well. I asked Huchra what he thought of both projects. "I wish them luck," he said.

We pulled into a small complex of buildings at the core of the VLA site. The place seemed utterly deserted. There was a telescope operator on duty, two astronomers with him, and no one else on the site that we were aware of. Hewitt drove us around, pointing out the massive building where the dishes are hauled for maintenance; what amounts to a railroad car with a hydraulic lift on it moves along a set of tracks, swings off on the siding where a single dish sits, slides underneath it and lifts it off its moorings. The car then trundles the entire telescope into the building. The tracks themselves, or at least the wooden ties they sit on, were secondhand when NRAO built the VLA, pulled up from defunct railroads and relocated to save money. They're now in such bad shape that it's considered risky to move the dishes at all along some sections of track, but the astronomers have to do it anyway to use the VLA at all.

Then, the short tour over the run itself not to start until the next afternoon and a full day of staring at computer screens in Socorro already behind them, Hewitt and Chen went back to work, reducing the unvarnished, noise-filled data from earlier runs into comprehensible images. For all astronomers, but especially for the calculation-intensive observations of radio-interferometric astronomers, this is the really grinding endurance work it takes to understand the universe.

The next morning, Hewitt was back at the computer again right after breakfast, but she took frequent side trips to the control room next door as the previous overnight observing run came to a close. In between it and her Einstein ring observations slated to begin by four P.M. was what the printed schedule described as "engineering time." That, she explained, was time built into the VLA's operations so software and hardware engineers could check out the telescope and make sure it was running properly. She told me that sometimes the checkout didn't interfere with taking data, although no one could ever predict that. So when the engineering time began, she waited until the operator looked like he had a free moment, and asked him what was going on. It turned out to be a software run, which would take only the first hour or so of the alloted six. "So, do you think we can do some observing when they're done?" she asked. "Sure, no problem." She had just gotten herself the most valuable commodity an astronomer can hope for: extra time on a major telescope. "And that," she said, "is one more reason I like being down here at the site. If I had been at the controls in Socorro, I would probably never have known about this, and the telescope would have sat idle all this time."

A radio run, for a visitor at least, is a lot duller than one on an optical telescope. There are no guide stars sliding across the video screen to look at; no images of perfect, miniature spiral galaxies piped in from the dome; no rumble as tons of glass and metal swivel to focus on the next object or as the building itself turns on its axis. I went outside for a walk, but because this run would be looking at the same object for hours, it wasn't even possible to see the dishes move; they stretched out in three directions until they seemed to disappear beyond the curve of the Earth. They were all moving in unison, but only fast enough to compensate for the turning of the Earth. When I went out later that evening, after the Sun had set, I didn't even have the luxury of seeing the stars. There was a light overcast, nowhere near enough to screen out the radio waves streaming in from the Einstein ring, but enough to make the skies look positively New Jerseyan.

The fact that the first gravitational lens distance measurement pointed to a universe old enough to contain the stars we can see is reassuring to cosmologists, but not all *that* reassuring. Not only is it based on a less than perfect understanding of the lens in 0957 but it's also contradicted by other new measurements. The results appear to be consistent, more or less, with each other, and they are converging on a figure that is uncomfortably high.

One of the observers who gets a higher value for the Hubble constant is Bob Kirshner, whom I had seen both in La Serena and in his Cambridge office. Mapping the universe is one of Kirshner's projects, but his real, original love was supernovas. "As a junior at Harvard, I did a research paper on the Crab Nebula," he told me on a return trip to Cambridge. Kirshner is now chairman of the department he majored in back in the late 1960s. "But don't make too much of that," he said. "The department is by far the smaller component of the Center for Astrophysics." The Crab is a glowing cloud of expanding gas left over from a supernova explosion that was chronicled by Chinese and European astronomers in 1054. "I did my graduate thesis at Caltech on supernovas, too, and this time I included my own observations of the Crab Nebula. When you start taking data that are better than anything that existed when you were an undergraduate, you start getting the feeling that you can really do something in this field. Anyway, the real reason I went to Kitt Peak and got involved in galaxy surveys was that it was the best way to be at the telescope for long stretches so I could be there when supernovas went off."

The galaxy work proved to be unexpectedly interesting, but Kirshner stayed interested in supernovas as well: he was heavily involved in studying SN 1987A, the exploding star that Tyson ended up observing. And although they aren't ideal as hundred-watt bulbs, Kirshner is using something called the Baade-Wesserling technique as another way of using supernovas to calibrate the relative distances of galaxies. Like the gravitational lens method, this one is an attempt to bypass the shaky distance ladder entirely.

The idea is based on old-fashioned trigonometry: if you know how big an object actually is and you also know how big it looks, you can figure out how far away it is. The objects Kirshner focuses on are the expanding clouds of incandescently hot gas blown off as a star explodes—the clouds that, hundreds of years later, thin out to look like the Crab Nebula. Paradoxically, it is much easier to figure out how big a supernova cloud actually is, even at a distance of tens of millions of light-years, than to figure out how big it looks from Earth. As the cloud expands at thousands of miles per hour, its light is blueshifted as it races toward Earth, the same phenomenon in reverse that accounts for the redshifting of galaxies' light as they race away. (Since the supernova's gas cloud expands more or less symmetrically, the far side of it is moving away from Earth and is redshifted, but that part is hidden from us). "We can use the blueshift to measure how fast the cloud is expanding quite accurately," said Kirshner, "which means we can say, in miles or kilometers, exactly how much bigger it is getting over any given interval of time."

Deciding how big it looks is a little more complicated, since a supernova cloud is far too small to resolve at extragalactic distance; it looks like a pinpoint of light no matter what telescope you use. But Kirshner has perfected a difficult technique, known as expansion parallax, that can be used to measure the invisible. "If the supernova were in perfect thermal equilibrium," he explained, "it would be a black body." That is, it would be something like the cosmic microwave background whose spectrum is uniquely determined by its temperature. "In practice, it isn't quite a black body, but we know enough about supernovas to adjust for the difference; you just have to solve a few simple equations." He held up a page covered with abstruse mathematical symbols. "See?" he said. "Easy. Okay, so you measure the spectrum and get the temperature. Once you have the temperature, you know how much energy is coming out of the cloud, not in total, but for every square centimeter of its surface."

Kirshner measures the energy output once, and then again about a week later. Over seven days, the apparent size has grown from A to B. He also knows, from the expansion rate, that the real size has grown from X to Y. The rest is mere algebra. "We tested the idea for 1987A," he said, "and found that we got the same number for the distance to the Large Magellanic Cloud that you get with other, independent measures. Over larger distances, we claim we're accurate to within twenty percent, which is a lot better than most." He showed me a chart. "If you plot redshift versus distance, you see that at bigger distances the redshift is greater. So we've discovered the expanding universe, although it's about sixty years too late to be a major discovery. Still, it's reassuring that our technique works. I don't claim we know the Hubble constant better than other people . . . yet. But the number we get is around sixty-five."

The third participant in the impromptu conference that had been held in the parking lot of the La Serena airport, along with Tyson and Kirshner, was John Tonry from MIT. Tonry has come up with a new way of measuring the relative distances of galaxies that tightens up the last, most uncertain step on the distance ladder. He is convinced that H-nought is even higher than Kirshner thinks. It would be worth mentioning that Tonry looks more like a kid in graduate school than a respected scientist (his trademark is an old daypack slung over one shoulder), except that by now I realize that this is the rule in astronomy rather than the notable exception. It's the astronomers who wear suits and ties who are remarkable. ("Usually the theorists," reads Ed Turner's marginal note.) There was no time to quiz Tonry in Chile, but I tracked him down a few months later at the American Astronomical Society's winter meeting, in January 1992, in Atlanta.

"It's really outstandingly interesting," he said, "that we're fi-

John Tonry

nally breaking a logjam after thirty years in really understanding the distance scale. It seems there really is a theoretical problem. This is something people have known about and ignored, because they could chalk it up to the uncertainty of the Hubble constant. That's not so easy anymore. There's been lots of new thought and lots of new observations, and it's all beginning to tie together. It looks as though the Hubble constant is about eighty. If omega really is one, then the actual age of the universe is eight billion years. That's really interesting because it's outrageous. There's going to be a lot of serious theoretical shakedown over the next decade, because we now know there's trouble."

Tonry's creative leap was in realizing that the science of statistics gave him a way to measure the apparent brightness of individual stars within distant galaxies thus enabling him to calculate the distance to those galaxies even though the stars can't possibly be seen as individuals. "Probably the best way to explain how I do this is by a kind of stupid analogy," he said. "Suppose you have a bunch of empty paint cans sitting outside, and there's a hailstorm. Then the Sun comes out, and all the hail melts. You look inside the cans, and there's water in them at different levels. The average amount of water in all the cans gives you a good idea of how much hail there was.

"What it doesn't tell you is how big the average hailstone was. Each can might have had hundreds of tiny stones, or a handful of big ones." But you can still get that information by looking not at the average amount of water but at the difference in water level from can to can. If the stones are large, the differences will be large, too—a couple of extra hailstones in one can will raise the water considerably above the average. If the stones are small, the differences will be small, too, since you can get a large number of extras without affecting the water level much. The amount of water from can to can is a clue to the size of the now-melted hailstones.

It works the same way with galaxies, which you can think of as buckets of stars. Tonry points his telescope at a galaxy and measures the amount of light falling into each pixel of a CCD. Then he uses the variation from one box to the next to calculate the apparent luminosity, as seen from Earth, of each individual star. He has, in effect, found a way to see, not hundred-watt light bulbs, but hundredth-of-a-watt light bulbs over enormous distances without actually seeing them. And he can use these dim bulbs to compare the relative distances of galaxies like M31, the Andromeda galaxy, whose distance is known to within 10 percent and galaxies like those in the Virgo cluster, which are far enough away that their redshifts are mostly out in the Hubble flow. (Out at that distance, the light entering each box on a CCD's grid comes from about ten thousand stars.)

"I check myself by using the technique to look at globular clusters within our own galaxy, whose distances are known, and it works. I've made observations of about two hundred galaxies with this method, and reduced the data on fifty or sixty. It takes a lot of work, a lot of careful analysis. But I'm getting a very good feel for this and a very good feel for its trustworthiness."

Tonry is not the only one who gets a Hubble constant that's uncomfortably high. Another bit of bad news comes from the good old Cepheid variable stars that led Hubble to the discovery of the expanding universe in the first place. At the same American Astronomical Society meeting where I cornered Tonry and got him to explain his distance-measurement method, two Caltech astronomers, Wendy Freedman and Barry Madore, announced that they had reexamined Cepheid stars in five nearby galaxies, Andromeda among them. Unlike Hubble and most of those who have monitored Cepheids since, they had not used visible-light detectors but CCDs sensitive to light on the edge between red and infrared light. That's an improvement in two ways. Infrared light can shine through clouds of dust undimmed, so that a Cepheid's brightness in those wavelengths is its real brightness, regardless of how dirty its home galaxy is. And the relationship between inherent brightness and pulsation period is firmer in the infrared than it is in visible light. The Cepheids still can't be seen out into the Hubble flow, but when Freedman and Madore combined their new numbers with the Tully-Fisher method of comparing the distances of spiral galaxies, they came up with an H-nought of eighty-five.

For all their close collaboration, Turner and Hewitt rarely see each other; the VLA run, had Turner made it, would have been an aberration. They remain closely connected through the telephone and especially through the electronic mail system, the same system that John Huchra used to find out about his bowling team's status by remote control from Mount Hopkins, and with which Tyson exchanged messages with his wife from Cerro Tololo.

But E-mail is no substitute for face-to-face meetings, and finally, two weeks after the VLA run, they would get to spend a week together. Joyce Turner comes from Newton, Massachusetts, near Boston, and the family was going to the area for a visit with the in-laws. Ed Turner would spend much of that time in Hewitt's office, helping her reduce data the two had taken over the years.

I arrived late in the week to watch them work together. It was my first visit to MIT, and it was a little bit overwhelming. Most of the buildings have numbers, not names, and the streets that crisscross the campus intersect at odd angles, as do the streets that border it. (This problem is not limited to MIT. It is a well-known navigational hazard throughout the greater Boston area. I once spent an hour trying to drive to a restaurant in downtown Boston. I was within a block of it for the last forty-five minutes and kept getting glimpses of it across vacant lots or the wrong way up one-way streets.) Hewitt's directions were just slightly ambiguous, and that was enough to add an extra twenty minutes to the walk from the Kendall Square subway exit to building 26, where she has her office. The building itself is drab and modern on the outside and utterly depressing on the inside. Hewitt's hallway looks like the corridor of a run-down inner-city high school: it stretches for what looks like hundreds of yards, flanked with gray cinderblock walls, dimly lit and with scuffed linoleum floors. The doors to most of the offices are windowless, closed, and locked. Hewitt's was, too, but she answered at the first knock and let me into a suddenly friendly environment: her office is spacious, with fourteen-foot-high ceilings, a desk, a table, and two comfortable armchairs. "Ed's in there," she said, pointing to an adjoining room, equally large, and so he was, hunched over a computer terminal and concentrating so intently he didn't notice us enter. Grace Chen was there as well, working at her own terminal.

One of the goals of the week was to get one set of observations in good enough shape to publish a paper on them, and Turner was in the middle of analyzing the visible-light images of one gravitational lens candidate. Finally he looked around and noticed us standing there. He looked a little bleary-eyed. "I've been Jackie's graduate student all week," he said. "She sure works them hard."

I asked him what he was doing. "Right now," he said, "I'm working on data that are decades old." This was a slight exaggeration, although the images he was analyzing were not exactly fresh. They were taken a few years earlier at Kitt Peak, and they showed a peculiar object that seemed to be a new lens. At the very center of his screen was a smudge amid the stars and galaxies, and under higher magnification the smudge resolved itself into three fuzzy blobs. "Here," said Hewitt, pulling out a sheet of paper. "Here are the radio plots of the same object. This is why we think it really is a lens." The radio plots were not pictures but maps of the intensity of radio energy. They looked much like the contour maps hikers use to scout terrain, with parallel lines marking out different levels of altitude. On radio maps, the lines mark levels of radio intensity. In this case, the map showed three peaks of emission, each looking something like a bull's-eye at the center of a target. In two different plots of the same object the three peaks looked pretty much the same—not necessarily the same in intensity but the same in overall structure.

"That is unusual," said Hewitt, "because the radio data were taken at two different wavelengths." If the source were a single object with several radio-bright components, the images should be different. The reason is that radio waves are typically emitted by electrons under acceleration; the three blobs of radio noise would presumably be moving at different speeds and thus each would be bright at a different wavelength. The fact that the pictures were so similar implied that there was a single, relatively compact region of electron acceleration, and that it had been split by a gravitational lens into three distinct images.

The optical image Turner had been staring at looked much the same as the radio images in terms of basic structure, which was another clue that this was a lens, not three separate objects. Just as with 0957, it was critical to understand just what the geometry of this system was—where the lensing object and the lensed object were in relation to each other and to Earth. Very subtle differences in the relative positions of the objects will cause differences in the brightnesses of each image, as well as in their apparent positions on the sky, so it is crucial to know just what the brightnesses are, and that was what Turner was trying to find out. He played the computer like some sort of video game: he would move the blinking cursor until it sat right over one of the three blobs, and then blot it out. Then he moved the cursor again and zapped the second image. What was left was a single image that the computer could analyze for light intensity. "It's tricky," he said, "because you have to worry about time delays." The same phenomenon that will ultimately reveal the size of the universe—the time delay— could screw up the measurements; if one of the images happened to be flaring during this particular exposure, that could throw all the calculations off. Another trick he would perform, once he had a good handle on just how bright the quasar images were, would be to subtract them from the picture entirely. That way, he could search for the intervening galaxy—the lensing object—whose precise location and shape would help them understand the geometry of the entire system.

Turner was not above complaining from time to time. "This image processing system is known as Vista," he told me. "It was created at Lick Observatory ten or fifteen years ago, and it's come down several evolutionary branches since then. This version is one that the radio astronomers use, and it doesn't work so well."

"No one likes us," said Hewitt.

"We like you," said Turner, "but you're not really astronomers. You're physicists. Not that that's anything to be too ashamed of." As an undergraduate at MIT, Turner had majored in physics.

"So, Ed, any discoveries?" she asked.

"Do you expect one every hour?"

It went on like that for most of the afternoon, with a break for lunch that involved a bewildering trip through a maze of interconnected buildings, a parking garage, and then a parking lot until, to my amazement, we came to a relatively normal-looking street with a normal-looking Indian restaurant on it. These are evidently two important components of the MIT experience: utterly ugly surroundings and plenty of good, spicy, exotic food (the exact opposite is true for Princeton).

Finally, at four forty-five in the afternoon, the image analysis had to stop. Hewitt was due at a wedding soon, and she and Turner hadn't yet gotten to the business at hand: organizing the paper they would write and submit to a scientific journal about their lens.

"Okay, maybe we should talk first about how we're going to illustrate it," said Hewitt. "What's the best optical image we have?" Turner handed her a black-and-white photograph of the lens. "Look at that!" she said. "There are fingerprints all over this!"

"I can't see them," said Turner, peering closely. "Anyway, the good news is that I finished the photometry. The bad news is that I've had no chance to digest it. But from these numbers"—he pointed to a list of figures—"we can get the instrumental colors. You can use those to check for relative magnitudes. But basically we have a hodgepodge. The only thing we can really address is the variation in these images, about two years apart. It didn't appear to me that there was any large variation, though. So . . . so what?"

"Well," said Hewitt, "what we want is a table of the apparent absolute magnitudes."

"That's not what they're called," said Turner. "Too bad they didn't educate you in graduate school."

"Well, whatever," said Hewitt. "I can see a table looking like this." She began sketching out a comparison of magnitudes for each of the images in different color bands over time. "So can you create these tables?" she asked him. "In a form suitable for a TeX file?" TeX, pronounced like "Tech," is a computer program that's used for typesetting technical papers and books.

"No," said Turner. "I don't do TeX. I do Troff [a comparable program]. I guess I'll have to learn it, though. It's an invention of the devil. What this object badly needs is a redshift. I wonder what we can actually say in this paper. The thing is that this object is sort of odd to begin with, so if it's lensed it would be nice if the oddness had to do with the lensing. I think we're missing something. How high does this object's redshift have to be to explain the fact that we don't see the lens?"

"Well, here's a calculation," said Hewitt. They pored over the figures.

"Okay, well, the lens, whatever it is, is very faint," said Turner. "And the source is very red. I guess I can see how we can put together a little data paper at least. I don't want to wait for more—I've learned you don't wait or someone else will do it." They looked at a picture of a spectrum. "Now, if that's the H-K break . . . ," said Turner. "If we believe this redshift, the lens itself would be *very* dark, consistent with being a totally dark lens. Basically," he explained to me, "you can't understand anything in astronomy without knowing how far away it is. We burned a *lot* of time on this doing spectroscopy. It would be hard to improve on it. I guess it's a Keck object." He was referring to the Keck telescope, still under construction at the time. "Well, I'll put together these tables," said Turner, "but we have to decide whether to make a big deal of it. Maybe it's a reddened quasar. Jerry [Ostriker] has always said we should be finding those."

"But it's not a quasar spectrum?" asked Hewitt.

"Well, if you assume this is the Lyman alpha break, the redshift is six point two." That would make the object by far the most distant thing, other than the microwave background, that astronomers have ever seen. "But, no," he continued, "it doesn't seem likely that it's a quasar. But it's got to be a lens. We can come down pretty hard on the lens interpretation. And on the fact that there's very little light from the lensing object. But we've got to get a spectrum. Maybe when I get back to Princeton there'll be a space telescope picture of this. There's no way of predicting, of course. The way things work with ST, you put your request in for an observation, but there's no way of knowing when they'll get to it. Sometime in the next year. Oh, well."

"Okay," said Hewitt, "we need charts and some speculation on this object."

"I can do that," said Turner. "I'm good at that kind of bull-shit—making ignorance seem profound."

"I'll work up the luminosities of the images," said Hewitt.

"Okay," said Turner, "I'll wait to do the bullshit until I know what to bull about."

"I still have to change clothes."

"All this would be a lot easier if I didn't have so much travel coming up."

"Well, when can you get to it?"

"Oh, about a month or two."

Hewitt looked offended. "Oh, come *on*, Ed!"

A month or two later, I was sitting in Turner's office at Princeton. On the floor was a thick reference book: *Atlas of Galaxies Useful for Measuring the Cosmological Distance Scale.* I asked him about the paper. "Well, we're working on it. Jackie's supposed to be writing it up. It's really the most fantastic thing I've seen in years. Very peculiar. There's really no good model for it. Jackie presented a version of the paper at a meeting on gravitational lenses in Hamburg last week, and it received fair attention. Now that we've gone public, I guess we're committed to doing something. Our conclusion at present is that the evidence is very persuasive that lensing is going on. The lensed object looks more or less ordinary in the radio, but the optical source is far too red and has no recognizable spectral features. One possibility is that we're looking at an ordinary object through a lot of dust, which would make it look redder. Another is that it's a normal object that's *very* far away, and that we're seeing some strange part of the spectrum that's been redshifted into the visible. That's sort of like being shown a photo of a person without being told that it was taken under a microscope. It wouldn't be surprising that it's hard to recognize.

"I've shown the spectrum to a lot of experts, and no one has seen anything like it before. It's fair to say we don't have a clue. It's a curiosity, and you never know where that will lead. I don't exactly know what more to do with it at this point." (By the following spring, infrared observations would confirm one of Turner's guesses: the object was probably a quasar seen through an uncommonly thick shroud of dust.)

Meanwhile, Hewitt had been reducing the data on her most promising candidate for a clean system with a time delay, the Einstein ring. Unfortunately, neither quasar image had flickered appreciably in all the times she'd looked at them. "It's frustrating," she said, "because we know that 0957 was quiescent for a while before it flared. The same thing could happen with the Einstein ring, and we could miss it. I can't get time at the VLA once a month on the chance that it will happen sooner or later. So now I'm thinking of trying to get time on a single dish to monitor the system regularly. You couldn't resolve the ring with one antenna, and so all I'd expect to see would be an overall increase in flux. But if I did, then I could run to VLA to look at it in detail."

CHAPTER SEVEN

NEW PHYSICS

Most astronomers feel the same way Ed Turner and Jackie Hewitt do about electronic mail: it's an effective way to stay in communication with colleagues anywhere in the world, to exchange news of clever theoretical ideas or remarkable observations or bowling-team scores, or to send drafts of papers back and forth so that two or more authors of a research paper can react to each others' contributions. But it's also no substitute for direct contact. No one has yet found a way to digitize body language, to capture on a floppy disk the subtle messages one human can send another with a slight emphasis of one word rather than another or of an eyebrow raised a few millimeters. That's why astrophysicists go to colloquia and seminars, why they travel hundreds of miles to attend the semiannual conferences held by the American Astronomical Society, and why, like George Blumenthal or David Schramm, they take the summer or the semester or even a sabbatical year to work at a different institution.

That is also why virtually every astronomer within a twenty-mile radius of Princeton tries to be at the Dillworth dining room at the Institute for Advanced Study on Tuesday at twelve-thirty P.M. for the ritual known as Tuesday Lunch. It is something like a colloquium, in that people talk about their very latest results, usually long before those results are made public, and answer questions about them; it is something like a conference, since on any given Tuesday there will be several speakers on topics ranging from dust in the Solar System to quantum fluctuations in the early universe; and it is something like a comedy improv club, because there is a premium placed on one-liners that bring down the house. Bob Kirshner would be an ideal participant.

The lunch is also something of a trial by fire. The host, unless he's in Washington testifying before Congress or consulting with NASA (both of which he does frequently) is John Bahcall, a professor of natural sciences at the institute, Neta Bahcall's husband, and an expert on, among other things, neutrinos. Bahcall sits at the center of the closed end of an enormous U-shaped table, usually dressed in a jogging suit and his Institute for Advanced Study baseball warmup jacket. Flanking Bahcall on most Tuesdays at this head table is Jeremiah (Jerry) Ostriker, head of astrophysics at the university, a mile or two to the north (although the two institutions cooperate in many ways, the institute is not and has never been affiliated with Princeton University). The others sitting at the bottom of the U are generally a visiting astronomer who has just finished giving the eleven A.M. colloquium in a nearby institute building, another visitor who will be giving a talk later on at the university, and two or three relatively senior local astronomers. Ed Turner often sits there. So does Lyman Spitzer, from the university, who was the first to come up with the idea of a space telescope back in the 1950s and who also did pioneering theoretical work on the idea of producing energy with controlled nuclear fusion, a project the Department of Energy is still pursuing to the tune of hundreds of millions of dollars every year. (Spitzer also had to be reprimanded by the dean of the faculty once during the 1970s, when he was well into middle age, for rock-climbing up the outside of the Gothic tower at Princeton's graduate college.)

   The astronomers and physicists filling up the sides of the U are no less impressive. Mixed in with the graduate students, postdocs, and junior faculty is John Wheeler, for example, whom Tony Tyson is convinced commutes by wormhole to be the maître d' at a dining room in La Serena. Wheeler is the man who invented the term *black hole* in the 1960s; he worked with such giants of twentieth-century physics as Niels Bohr and Albert Einstein. Freeman Dyson, who worked on the Manhattan project, helped design a rocket propelled by atomic bombs (it was never built, though a conventional-explosives model was) and wrote an influential antinuclear book in the 1980s, is usually there. So is Fang Lizhi, the dissident Chinese astrophysicist. On his escape from China, he more or less went straight to the institute for a year as a visiting fellow. Also there are Jim Gunn and Don Schneider, at the university and the institute, respectively, who share with their collaborator Maarten Schmidt the record for the most distant object ever discovered, a quasar with a redshift of 4.9 (it was announced at Tuesday Lunch long before the public got wind of it). Jim Peebles, of the Harrison-Zel'dovich-Peebles spectrum, the early papers on cold dark matter and the original Dicke team that went out to look for the cosmic microwave background, often shows up. So does Bob Wilson, who, along with Arno Penzias, found it and shared the Nobel Prize. Joe Taylor from the university, who is probably the world expert on pulsars, also comes as do Dave Wilkinson, Tony Tyson, and Lyman Page. With the combined resources of Princeton University, the institute, Bell Labs, and Rutgers University, the Tuesday Lunch is an extraordinary concentration of astrophysical talent.
   What makes it a trial by fire is the way John Bahcall runs it. After he calls the group to order by banging on his drinking glass with a knife (he also uses it to hush people who start independent conversations), he usually calls on the speaker who has just finished giving the institute colloquium, and then on the one who will be giving the one at the university later. Instead of having them discuss their colloquium topics, he insists that they talk about

something else they're working on, or that someone else at their home institutions is working on. Bahcall will also call on one or two locals, and if they're lucky he will have given them a day or two warning. Sometimes he does not. Bohdan Paczynski, from the university, was standing in the cafeteria line waiting to get lunch one Tuesday when Bahcall stepped up and informed him that he would be tapped that day. "I hope you have something interesting to talk about," he said.

As it happened, Paczynski did. He had been thinking about the problem of gamma ray bursts, mysterious flashes of high-energy electromagnetic radiation, which were first detected in the 1970s and periodically pop off in the sky for reasons no one yet understands. The presumption at the time was that they came from some sort of violent process in the Milky Way—matter crashing down onto the surface of a superdense neutron star, perhaps. "I was wondering," he told me later, in a thick Polish accent, "of something altogether crazy. Maybe these gamma ray bursts are cosmological, coming from outside the Milky Way. How would we determine this? Well, if it turned out that these bursts were isotropic, coming from all directions, that would be strong evidence. The only things we know of that surround us uniformly in all directions are the Oort Cloud of comets that surround the Solar System beyond the orbit of Pluto and the universe. It would be really crazy to say that bursts come from the Oort Cloud, since we have no way to produce them. So maybe it is slightly less crazy to think that they are cosmological."

It was still crazy, though, and that made it seem just right coming from Paczynski. He is a highly respected theorist, but he looks like a mad scientist from the 1950s. He wears his hair in a severe crew cut, sports eyeglasses with thick black plastic rims, and will say things like "Over next few days I will be traveling, so I cannot say what my coordinates will be on Thursday," or "Between ten and eleven tomorrow morning, probability is ninety-five percent that I will be in my office."

Paczynski championed his crazy idea, first aired at Tuesday

Bohdan Paczynski

PHOTO: EILEEN HOHMUTH-LEMONICK

Lunch, for several years afterwards. In 1991, NASA's Compton Gamma Ray Observatory satellite, the first probe sensitive enough to test his idea, found that the gamma ray bursts are indeed isotropic. They may originate when a neutron star spirals into a black hole and is vaporized. Soon thereafter Martin Rees from Cambridge, a coauthor with Blumenthal, Primack, and Faber on the 1984 paper that convinced the world that cold dark matter should be taken seriously, was in town for the institute colloquium and arrived for Tuesday Lunch. Bahcall called on him and began to explain the ground rules. "I've been to this lunch," Rees interrupted. "I know the rules. Before I discuss something that I haven't covered in my talk, I have a story that will make both me and Bohdan look foolish. A year ago, I offered him odds of one hundred to one that the gamma ray bursts were inside the galaxy." He paused. "And he didn't take me up on it." The room broke up.

Another crazy idea that came out of Tuesday Lunch one day

was J. Richard Gott's time machine. Gott is a theorist at the university, a boyishly pudgy-faced Kentuckian who never lost his native accent. He has the air of a good old boy from the hills. You can just imagine him sitting on the porch of the general store, swapping stories with the locals. He is a natural born storyteller. As he talks, he gestures constantly, and his voice rises and falls in both pitch and volume to contribute to the dramatic effect. It's just that instead of talking about hunting 'possums and tracking 'coons, Gott talks about N-body simulations and general relativity. He is one of the last great defenders of standard cold dark matter and maintains that the simulations he runs with a graduate student, Changbom Park, prove that CDM is in perfect health. "See, look right here," he said one day after lunch, pointing at a sheet of paper covered with tiny dots. "See, there's the Great Wall, there are the voids—I just don't see any problem."

Gott's time machine consists of three objects: a spaceship and two cosmic strings. Like inflation, strings are a consequence of the

J. Richard Gott   PHOTO: EILEEN HOHMUTH-LEMONICK

phase transitions that may have taken place in the early universe when the underlying energy fields of the cosmos cooled and, like water freezing into ice, changed in form while staying the same in substance. When a pond freezes, the ice begins forming at several places at once, condensing around an impurity in the water and spreading outward. As the water crystallizes, its molecules all line up in the same direction, but there's no reason that direction should be the same for each center of condensation. It isn't, in fact, and when two growing ice blocks meet, they can't line up. There is a discontinuity, which shows up as a line or a sheet of cloudiness in an otherwise clear medium.

Or take the analogy preferred by David Spergel, a young Princeton astrophysicist and regular Tuesday Luncher who lives in New York City and commutes to Princeton by train (his wife is in medical school, so it's more convenient—and also, in the mind of a native New Yorker, a lot more interesting—to live in Manhattan than in stodgy Princeton). "Think of two nearby cities that start building grids of roads," he told me one day after lunch. Within each city, the roads are regularly spaced and at right angles, like the streets and avenues of midtown Manhattan. But the grids are pointed in different directions. The avenues in the first city point, say, north by northwest, while those in the second city point directly north. No one notices a problem until the cities begin to grow. Eventually, their road systems will meet. Since they are not oriented the same way, the streets will have to bend if they are to join up. The perfect regularity of the grid patterns will now be marred by a defect, a place where straight roads kink.

A cosmic string is analogous to the boundary between two disoriented chunks of ice or two road networks. It marks an area where two sections of the newborn universe couldn't line up as they condensed into a new energy phase. One consequence is that the strings themselves, unable to relax into the low-energy state of the modern universe, preserve the extraordinary density and energy of its earliest fractions of a second. A cosmic string, which is thinner than an atomic nucleus, can stretch all the way across the visible universe and has a mass of billions of tons *per inch*. It goes without saying among astrophysicists that while cosmic strings are fascinating objects to theorize about, no one has ever observed one, and while they may exist in principle there is no guarantee that they exist for real.

But if they did exist, Gott explained, then they would be massive enough to warp spacetime to a fare-thee-well. And if a spaceship managed to make a circuit once around this pair of fast-moving strings, "it would follow a closed timelike curve. Someone on the ship would be able to visit his own past. Here, I've brought some visual aids." He reached into his sportcoat pocket and pulled out two lengths of red yarn and a toy space shuttle. Then he dangled the two cosmic strings and made the ship go once around. "Of course, I'm only saying this would work in principle. But it does raise some interesting questions about causality—after all, if you could really go into the past and kill your grandma before you were born, physics will have to find some way to deal with that."

Since then, other astrophysicists and pure physicists have attacked Gott's idea, claiming his analysis was incomplete, while others have supported it. Gott himself has written a paper to refute the critics in a series of articles published in the prestigious *Physical Review Letters*. But whether the time machine turns out to be a real phenomenon, a viable but purely theoretical concept, or all a big mistake, it will have gotten astrophysicists thinking along new lines. That is valuable in itself. When scientists confront and wrestle with crazy ideas, they get an intellectual workout at a minimum, and sometimes, as in the case of Paczynski's extragalactic gamma ray bursts, they even come up with unexpected solutions to serious puzzles. Generally speaking, the more serious the puzzle is and the longer it goes unsolved, the crazier the proposed explanations become.

That has proven to be true for the observational and theoretical crisis facing cosmology. As observers like Huchra and Faber came up with Great Walls and Great Attractors; as Don Schneider and Jim Gunn found quasars, and hence the earlier and earlier existence of cosmic structure; and as the COBE satellite probed deeper into the light from the edge of the universe without finding any ripples, theorists began scrambling to squeeze very large or very early cosmic objects out of their equations, objects that the standard cold dark matter model could not easily explain.

Cold dark matter itself is a kind of theoretical rabbit, an elaboration of the original Big Bang model of the universe pulled from the hats of theorists. "Remember," George Blumenthal told me out in Santa Cruz, "that CDM was originally invented because we should have seen microwave background fluctuations at the level of one part in a thousand or one part in ten thousand, and we didn't." The original Big Bang was widely accepted because it explained one well-known phenomenon—the expanding universe—that otherwise couldn't be explained, and because it made two specific and unique predictions—that there should be a cosmic microwave background radiation and that the ratio of hydrogen to helium to lithium should be 75 to 25 to .00000001, which later turned out to be correct. CDM, with all its assumptions, was accepted as a modification because it explained where the original fluctuations had come from, how galaxies could begin forming at the right mass scales, without leaving an imprint on the microwave background, and how the cosmos could look the same in all directions. But its predictions hadn't been tested yet: no one had seen the mass needed to make omega equal to one; no one had seen the microwave background fluctuations CDM demanded; no one had found the particles that people like Blumenthal and his partner Joel Primack had suggested is the stuff dark matter is made of. Now, anticipating that none of these things might be found, it was time for more rabbit pulling; perhaps CDM needed some more modification, or perhaps it had to be thrown out altogether and replaced with a new theory, incorporating, as CDM had in the early eighties, some sort of new physics.

One tiny group of scientists even argued the cosmological crisis was so deep that the most fundamental concept in modern astrophysics should be thrown out. They felt the Big Bang itself was bogus. The universe has existed for hundreds of billions of years, at least. The galaxies and clusters of galaxies did not form under the influence of gravity but rather by electromagnetism, acting on the hot, electrically charged gases known as plasmas.

These plasma cosmologists are followers of Hannes Alfvén, a Swedish physicist who won the Nobel Prize in 1970 for his theoretical analysis of the nearest major blob of plasma, the Sun. Their argument, set forth in scientific papers and in the popular book *The Big Bang Never Happened* (written by a free-lance cosmologist named Eric Lerner, who lives and works about five miles south of Tuesday Lunch), is that the most natural way to understand the universe is to study what happens in terrestrial laboratories, then generalize. Thus, since plasma confined in, say, a neon light fixture will twist itself into electrically charged sheets and filaments when subjected to an electric field, the plasma out in the universe should do the same. That, they say, explains the long filaments of galaxies that Geller and Huchra see in their surveys. It also explains the so-called cosmic microwave background: there was no primordial fireball of a Big Bang, but rather a network of plasma filaments pervading the universe, acting as gigantic attenuators that absorb and reradiate radio waves coming from active galaxies. Redshifts don't arise from the recession of galaxies in an expanding universe, but from bizarre effects seen in some tabletop experiments.

Leaving aside the dubious assumption that the universe mirrors the laboratory (try setting two bowling balls close together on a table and see whether they crash together as two planets would, under their mutual gravity), most conventional cosmologists regard plasma cosmology with some amusement. They acknowledge that plasma forces may have a significant role in the evolution of galaxies, but the idea that the inconsistencies in the Big Bang model and in CDM should be replaced by a universe that could have been designed by Rube Goldberg strikes them as ridiculous. The suggestion made by some plasmologists that astrophysicists are part of an intellectual conspiracy to suppress inconvenient data makes them a little bit annoyed. "If I thought this model had the slightest chance of replacing the Big Bang, I would immediately begin working on it, since it could lead to the Nobel Prize," Bohdan Paczynski once told me. "It's absurd to think I would try and repress such ideas."

In fact, the Big Bang may well be wrong, and may someday be replaced by another model. Limited shelf life is part of the territory for models of the universe, which are nothing more than descriptions of Nature that best fit the facts at a given time. Isaac Newton's theory of gravity was breathtakingly successful; it predicted at once the behavior of falling objects and of the orbiting planets with such accuracy that NASA still uses it to calculate the trajectory of its spacecraft. It went unchallenged for more than two centuries until Albert Einstein created the general theory of relativity, which equaled Newton's theory in predicting motion in ordinary situations but exceeded it in describing the movements of bodies at very high speeds or in the vicinity of very massive objects. When and if the Big Bang is finally replaced, its replacement will have to do comparably well.

Although the plasma cosmologists haven't yet made their appearance at Tuesday Lunch, there is plenty of rabbit-pulling that goes on. One day, for example, Jerry Ostriker presented his latest idea: there really isn't any nonbaryonic dark matter at all. This is the same Ostriker whose papers in the early 1970s with Jim Peebles convinced many astronomers to take dark matter seriously after decades of neglect. Back then, according to Margaret Geller, who was in graduate school at Princeton at the time, "Jerry was considered the young Turk of the department." In some ways, that is hard to imagine: Ostriker is shorter than average, balding on top, and he squints when he's trying to puzzle out an idea. But though he has aged physically, he hasn't lost a taste for iconoclastic ideas.

One, which he coauthored with institute physicist Ed Witten a few years ago, argued that the cosmic strings Richard Gott was using for his time machine might have given structure to the universe. According to Witten's calculations, the strings could be

Jeremiah Ostriker   PHOTO: EILEEN HOHMUTH-LEMONICK

superconductors: they could carry electric current without resistance. If there were some sort of magnetic fields in the early universe (and no one knows why there should have been) then these
would have induced electric currents in the superconducting
strings. Surging with electricity, loops of string would have acted
as gigantic radio transmitters, sweeping up matter over enormous
volumes of space and piling it into spherical shells. The fact that

redshift surveys by Geller, Huchra, and Kirshner have found just that kind of structure would have been powerful evidence for the idea, except that the structures are much too big. Anyway, early universe radio broadcasts would have distorted the microwave background radiation in ways that COBE should have seen by now. Superconducting cosmic strings are no longer taken seriously.

But now Ostriker was ready with his new idea. "Okay," he said, having gotten the floor from John Bahcall, "you really have two problems with dark matter. The first one is that the nucleosynthesis arguments tell you that you can only get to about five percent of critical density in baryons, but you can measure dynamically that galaxies and clusters have about ten to twenty percent of critical density. So you assume there's some kind of nonbaryonic dark matter. Then, if you want to go all the way to omega equals one, you have to add something else, which you *don't* measure dynamically. It's really a zoo. I think that there's something fundamentally wrong, and I'll put it simply: omega is not equal to one in dark matter. The simplest argument for that, which I know from the early work I did on dynamics, is that when you have an omega of one you have steady merging of systems."

His argument was that if the universe is at critical density, there have to be some places where the density is slightly higher than average. If you draw a circle around two such regions, you've essentially described a miniuniverse where omega is a little greater than one. The miniuniverse must therefore collapse; the two regions will merge into one. Now draw a circle around that new, doubly overdense region and the nearest one like it; once again, the two must merge. "You can't stop it," he said. "Now, do we see very frequent mergers of galaxies in the universe? No. Think about the Milky Way and suppose something came spiraling into it. It would heat up our galaxy's disk and disrupt it. You can show that in order to keep the Milky Way as thin as it is, no more than three percent of its mass could have come in since the stars were made.

So what does this mean? If our galaxy is typical, then models of structure where omega is one can be ruled out. That's all. I think this is the strongest argument.

"Okay, so we're saying this: let's get rid of the extra, and say that what we measure dynamically is not some exotic dark matter, just baryons. How do you account for the light-element abundances? Well, remember that the Jeans mass at decoupling is ten to the seventh solar masses. Okay, what are the things that would form at this mass? Black holes. Most of the matter would collapse to stars and collapse to black holes. Not all, because as you put more and more into black holes, the density of the rest gets less and less. It gets harder and harder to pull more in. So it's not incredible that ninety percent of the baryons would have gone into black holes. Now you have them floating around in this sea of gas. What happens? The black holes form accretion disks [these are disk-shaped clouds formed by gas piling up as it tries to fall into black holes; the compressed gas heats up to incandescence before it disappears]. What does an accretion disk do? Emits high-energy radiation, like the quasars. So we let the radiation from black holes go out through the surrounding gunk. What does it do? Breaks up helium four, gives you more deuterium, lithium, changes the abundances, and makes it look like there's less baryonic matter. We've now accounted for the dynamic dark matter with baryons."

"Where are the black holes now?" someone asked.

"They're still here, in the halo of the galaxy."

Well, maybe. As soon as Ostriker was done, the questions began to fly. "How do you account for the flatness problem if omega is only point two?" "How would you go about detecting these black holes?" "What about the microwave background?" It was clear that the rest of the lunchers were nowhere near ready to accept the latest Ostrikerism. "I'm just saying it's plausible," he protested.

Another new idea that periodically bubbled up at Tuesday Lunch is Jim Peebles' latest theory. Having abandoned the CDM model he helped to create, Peebles, ever the contrarian, was now

P. James E. Peebles

PHOTO: EILEEN HOHMUTH-LEMONICK

P.J.E. Peebles *(left)* talking with Neil Turok *(right)* at the Princeton COBE workshop

PHOTO: MICHAEL D. LEMONICK

pushing something called PIB, for Primordial Isocurvature Baryons (though some astrophysicists insist the P originally stood for Peebles). "Actually," he told me during a visit to his Jadwin Hall office, "I'm going to change the name to Baryonic Dark Matter. It has a better ring, don't you think?" As he explained PIB/BDM to me, the tall, skinny Peebles leaned back in his chair, his long legs propped up on the desk, and fiddled with a toy helicopter someone had given him. The conversation was punctuated from time to time by the whir of its blades as the toy rose from his hands and sailed across the room. "I'm very enthusiastic about this model," he said, "but then, it's inevitable in this business that you think whatever you're working on is the greatest idea you ever had. Otherwise, you'd lose interest rather quickly."

The beauty of PIB is that, like Ostriker's model, it requires no exotic particles at all. The dark matter responsible for making the galaxies move too fast is made of ordinary baryons, says the theory. These baryons were clumped together in proto-clusters long before matter and energy decoupled from each other. That would have left a strong imprint on the microwave background—

except that in PIB, every place in the early universe that was formed with an excess of matter was also formed with a deficit of radiation. The cumulative effect on the microwave background: zero.

The ugliness of PIB, though, which Peebles freely admitted, is that, unlike CDM, it incorporates no good explanation for why the fluctuations in matter density follow the pattern they do. Moreover, the theory is imprecise enough that it doesn't lend itself to straightforward calculations. It is therefore hard to use it for making concrete, testable predictions.

Another idea that made the rounds a few years ago was that the universe's largest structures could have come from cosmic strings, not as radio beacons sweeping up matter but as massive seeds, serving the same function as clumps of cold dark matter particles by gravitationally gathering baryons by the trillions of solar masses to form galaxies, clusters, and superclusters. That theory is already on the wane, the victim of supercomputer simulations, but one of its strongest advocates is still around working in the Princeton physics department. He has given up on strings, but his latest idea, presented in due course at Tuesday Lunch, offered still another way to build a universe.

When I met Neil Turok in his office in Jadwin Hall, I mistook him at first for a graduate student waiting to see the eminent theorist Neil Turok. He looks a good half decade younger than his thirty-two years, and while he is rather shy and casual, he is considered by his colleagues to be an uncommonly brilliant theorist. Turok was born in South Africa, the son of a land surveyor and his wife who, fed up with apartheid, went into liberal politics. This was not a healthy hobby in the South Africa of the 1960s, and the family was forced to flee the country. They lived in East Africa for four years, then went on to London, where Neil became a British citizen. His parents were finally given permission to return to their country and, encouraged by the political changes there, have done so. Neil has been back only once, between college and graduate school. He taught for a year at an elementary school in the quasi-independent black homeland of Lesotho.

Originally, Turok was not planning to be a physicist at all. "I was interested in biology as an undergraduate at Oxford," he said, "specifically in evolutionary biology and ecology. But after a year, I decided that biology was just not precise enough a science for me, and I switched to physics. I thought it was the best thing around, and I still do. It's very hard-nosed, very practical—you have to keep testing your theory with experiment—but it also addresses the very biggest questions." He ended up majoring in theoretical physics and went on to graduate study at Imperial College in London. "My thesis wasn't particularly coherent," he admits. "It was more of an anthology. Part one was about cosmology, but parts two and three were about field theory."

After he took his Ph.D., Turok did a year of postdoctoral study at the Institute for Theoretical Physics at the University of California, Santa Barbara. "It was there that I went to a workshop on cosmology, and this was just at the point where inflation was taking off. I realized that cosmology was something I wanted to focus on." One of the first things Turok did was to help write a computer code that could be used to model the evolution of cosmic strings in an expanding universe. After Santa Barbara, he went back to Imperial College for two years, but the job prospects were meager and the morale was low. So when he got an offer to go to Fermilab, where David Schramm was working, he took it and then moved to Princeton when another offer came along the next year.

"I had been at Princeton for about a year working on cosmic strings," he said, "and getting kind of frustrated. The theory seemed to get more and more complicated the closer I looked at it, and it was very hard to make calculations. Also, it didn't look very promising from an observational point of view. For one thing, to make the kind of large-scale structure we see, you need to make bubbles at a scale of thirty megaparsecs. The string network you see coming out of the calculations has a smaller scale than that. I was discouraged when it didn't work out. It seemed that I had been wasting my time."

Turok was still interested in cosmology, though, and he began trying to think of other sorts of exotic physics that might solve the problem of structure formation. "I was wrestling very hard with the problem," he said, "and I wasn't getting anywhere. Finally, I gave up and went on holiday to Egypt to unwind." He was in a plane flying over the pyramids when a thought struck him: "defects with a third homotopy group."

The phrase would have been meaningless even to most physicists, but to Turok it was a revelation. It meant that there might finally be a way to explain the structure of the universe. Physicists who play with the ideas of phase transitions in the early universe know that a cosmic string is only one of the kinds of defect that can show up in an imperfectly congealing energy field. Another sort is the monopole, the particle Alan Guth had been trying to understand when he stumbled on inflation. Another is the domain wall, a sheet of uncondensed energy that is the two-dimensional counterpart of a one-dimensional string. And a fourth kind, as Turok realized while overflying the pyramids, is a defect with a third homotopy group, also known as a texture.

There is no easy way to describe textures, except perhaps as knots in the energy fields of the universe. They are not entirely new to physics. Turok explained that, in fact, the proton, a basic building block of atomic nuclei and the nucleus of the hydrogen atom, can be thought of as a texture within the electroweak field. "And that," he said, somewhat defensively, "is very conservative physics."

But cosmological textures, as opposed to subatomic ones, were different things entirely. Back in 1977, a physicist named Tom Kibble, whom Turok studied with at Imperial College, was exploring the idea that phase transitions had played a part in the early universe, and he briefly looked at the idea of texture. "It is hard to envisage any possible physical relevance," wrote Kibble at the time. "He's kind of embarrassed by that statement now," said

Turok with a grin. What Kibble meant was that textures, if they existed, would tend to be as large as the horizon scale, as large as the visible universe, and thus not detectable because there wouldn't be anything to compare them with.

Turok's leap of insight was to realize that while textures in the modern universe would be undetectable, there should have been textures in the early universe as well. They, too, should have been as large as the horizon scale, but that scale was a lot smaller ten or twenty billion years ago than it is now. As the universe expanded, the imprint on matter left behind by the textures at that epoch should have persisted to the present. "The structures we see now, the bubbles in redshift surveys," explained Turok, "are on the order of thirty megaparsecs across. That, as it turns out, is just about the size of the universe itself at the time that matter and radiation decoupled, the time when the microwave background began to shine."

Because a texture is a knot of higher than average energy, it is also a knot of higher than average gravity, and matter should be drawn into it. There, at a stroke, might be an explanation for the large-scale structure of the universe. If matter was already beginning to collect at the thirty-megaparsec scale at a redshift of one thousand, it is easy to imagine how quasars could have formed soon thereafter and how Great Walls and Great Attractors were born. "The fact that the biggest structures are just the same as the horizon scale at decoupling is an interesting coincidence, at the very least," said Turok. "What attracts me so much is that this theory is *very* falsifiable. If we found structure at a level of three hundred megaparsecs, for example, that would quickly kill it. The theory is also very calculable.

"It's gone from a vague idea in 1989 that you can get horizon-scale physics to the remarkable idea that this might explain structure easily. Inflation has been very popular for a long time, but it still depends on a lot of fine-tuning the conditions, and it doesn't say anything about what happened *before* the inflationary period. It does make predictions, though, about how the microwave background radiation should look, and those are close to being ruled out—in fact, they may already be ruled out."

Although inflation is a theory that competed with his own, Turok was unwilling to kill it prematurely. "You don't want to do that," he said, "because we don't have many good theories to begin with." Simplicity appeals to him. "The reason I work on this stuff," he said, "is that while there's still a chance that these simple theories will really explain the universe, I think that's well worth pursuing. If the universe turns out to be very complicated, I'll quite probably move on to something else."

There was already some preliminary evidence that adding texture to the existing cold dark matter model of the universe could explain most or even all of the problems of large-scale structure. Turok had gotten together with Jerry Ostriker, David Spergel, and a graduate student, R. Y. Cen, to concoct a massive simulation of the universe. Their model began with the notion that cold dark matter does indeed exist, along with ordinary matter, and that both live in a cosmos dominated by gravity.

But then they added two crucial features: one was texture, planted as a seed to see if it could grow galaxies and large-scale structures. And the other was hydrodynamics, the laws of physics that govern the behavior of gases and other fluids. "You have to remember," said Ostriker, "that initially there were no galaxies, just gas. That gas had to cool and collapse, and fragment and make itself into galaxies. That process is hydrodynamics. Most of the codes astrophysicists use in their simulations presume that the universe is made of pure dark matter. That's fine if we could observe dark matter directly. But since we're observing galaxies, then you'd better compute when and where the galaxies form."

Basically, the simulation worked. It created an electronic universe that resembles the real one. It had early-forming quasars, large-scale flows, large structures, and giant voids: all the features Turok had outlined for me as the basic test of a viable theory. "Let's put it this way," he said. "Even if this theory is wrong, it gives so many remarkably good answers that it can't be *entirely* wrong."

A successful simulation is far from proof, however, Turok emphasized. "We have to be checked by other groups; they have to get the same results we do. And like any useful model, we have to make predictions that can be tested by observations. The strongest prediction we're making is on the microwave background: textures in the early universe should have left a pattern of hot spots in the microwave sky unlike the pattern predicted by anyone else." COBE should be able to detect this pattern, and if it's not there, the theory in its simplest and most believable form, will be proven false.

Another sort of new physics that could resolve the crisis is actually very old: Albert Einstein invented it but later retracted it, calling the idea the biggest blunder of his life. When Einstein used the equations of his new general theory of relativity to model the behavior of the entire universe, he found that the cosmos he had constructed was unstable. It had to be either expanding or collapsing; what it could not do for any length of time was stand still. Yet there was no evidence at the time that the universe was doing anything else. The stars and galaxies might be moving, albeit so slowly that the motion was imperceptible, but surely these motions canceled out when averaged over large distances. The universe as a whole was stagnant. To reconcile the equations with the real universe, Einstein introduced a "cosmological term," an extra term in the equations that had the effect of keeping things steady. It amounted to an undetected form of energy that pervaded the universe, generating a repulsive force precisely equal to but opposite that of gravity, which would keep the cosmos inflated, like hot air puffs out a balloon.

Then along came Hubble and his demonstration that the universe was in fact expanding. If Einstein had left his beautiful equations alone, he might be remembered, along with everything else he did, for predicting the expanding universe. Hubble's discovery, along with other data led Einstein to renounce the constant. Since then, though, it has kept sneaking back into physics and astrophysics, when the scientists least expect it. As Ed Turner and Bill Press wrote in a review paper they coauthored with Sean Carroll at Harvard, "The cosmological constant is an idea whose time has come . . . and gone . . . and come . . . and so on." I asked Ed Turner whom to talk to about the cosmological constant, and he referred me to the unrelated Michael Turner, at Fermilab, David Schramm's longtime colleague and the author of a paper pointing out the advantages of the cosmological constant for solving astrophysics' many problems.

I called him and mentioned Ed Turner's name. "Ah," he said, "my evil twin." Michael's hair may be thinning, like Ed's, and their last names may be the same, but there the comparison ends. Ed is round; Michael is rangy. Ed is soft-spoken and reserved; Michael's voice is strong, and he does not shy away from strong expressions of opinion. He is also something of a showman; the

Michael Turner
PHOTO: LAUREN SHAY

viewgraphs he displays at conferences are legendary (one showed the entire history of the universe on a single page, with characters from the Pac-Man video game representing various elementary particles), and he was once photographed for a magazine article wearing a novelty headband with two Styrofoam balls waving at the end of antenna-like springs. Just as David Schramm had done when I was in California, Michael Turner saved me a trip to Chicago: he was coming to Princeton for just a day to give a talk at the institute. Unfortunately, he had no spare time so I offered to pick him up at Philadelphia airport, and ended up speaking to him, tape recorder in one hand, as we drove north on Interstate 95.

"The cosmological constant is this crazy thing, right?" he asked. "Throughout the ages, astronomers have used it to save themselves from crises. Einstein started it—he put in a cosmological constant to make his expanding universe solutions not expand. You could say Einstein was stupid to put it in, which he himself said. But in the context of modern theory, it's quite the reverse. That term has every right to be there. It has as much right to be in the equations as any other term. Murray Gell-Mann [a Nobel Prize winner from Caltech, who was among the first to realize that subatomic particles are made from even smaller particles, which he called quarks] is often quoted as saying that anything that's not forbidden by physics is mandatory, and I think that's more true than false. There's no principle that forbids a cosmological constant. So the question isn't why did Einstein put it in—it should have been in from the beginning."

There is reason to expect a cosmological constant because of quantum field theory, and in particular because of the same Heisenberg uncertainty principle that gave rise to the quantum fluctuations which happen during inflation. The uncertainty principle allows pairs of particles—an ordinary particle plus its corresponding antiparticle—to spring into being spontaneously, out of literally nothing, as long as they annihilate each other (as particles and antiparticles always do when they come into contact) almost immediately. These "virtual particles" pop into and out of existence so fast that they can't be detected, even in principle, but there so many of them popping throughout empty space all the time that they charge the vacuum with energy.

As Turner explained, the trouble is that the amount of energy these particles should generate, according to standard particle theory, is enormous. "Basically," he said, "it should be about one hundred twenty-two orders of magnitude bigger than we know it is." That's one followed by 122 zeros. And where Einstein's constant was designed to just hold the universe up against collapse, the particle-physics cosmological constant should be making the universe fly apart 1,000,000,000,000,000,000,000,000,000,000,000, 000,000,000,000,000,000,000,000,000 times faster than it is.

"So already you can see that the cosmological constant is so bizarre that you can't even call it ugly. You call something ugly in physics when it violates some cherished belief about the universe, but we have so little understanding of this thing that you can't even say that. The thing is, though, that if you somehow sweep this enormous number under the rug, and give the constant some reasonable value, it does a number of things for you. It resolves the age problem. It pushes more power to large scales; it makes the simulations fit the real universe like a glove. It solves the omega problem. And finally—I'm a used-car salesman, obviously—it does all this without pushing the fluctuations you'd expect to see in the microwave background up by very much, so COBE hasn't yet ruled it out."

The way it does all this is by filling the universe with a uniform bath of energy. Because energy generates gravity just like matter does (Einstein showed that matter and energy are equivalent), a cosmological constant could make up the difference between the observed density of matter (an omega of .2, based on the motions of galaxies in clusters) and the critical density (an omega of 1), which theorists favor. There could be enough of a cosmological constant, in other words, to account for .8 of the critical density, yet because the constant is perfectly uniform through space, it would pull individual galaxies in all directions at once, and thus be undetectable. Because the energy of the cosmological constant tends to inflate the universe and because its influence grows as the universe expands, the universe may actually be expanding faster now than it was in the past, not slowing down. That makes the universe older than its present expansion speed would imply, which gives plenty of time for the oldest stars to have lived their lives. And finally, on the very large scales represented by the Great Wall and the Great Attractor, a cosmological constant would accentuate the tendency for large structures to grow, bringing these objects within the bounds of CDM.

The cosmological constant is plausible, ugly, and useful. It is also testable by a means Ed Turner came up with. Because the universe would be speeding up under the influence of a cosmological constant, the relationship between redshift and distance would be much harder to calculate. The farthest objects—the quasars, for example—would really be farther away than they seem, and there would be more galaxies between here and there to get in their way and act as lenses. If you look at a large number of quasars with a high-resolution telescope like the Hubble space telescope (even in its current blurred-out condition), you should see more lenses if there is a cosmological constant than if there isn't. Another useful test, based on the same principle, is to count the number of galaxies in comparable volumes of space near and far; if the universe is deeper than it looks, there will be more galaxies in the faraway volume, on average, than close by. (The galaxies have to be comparable, though. Tony Tyson's faint blue galaxies are almost certainly different from modern galaxies; the fact that they're more numerous doesn't prove anything.)

"The gravitational lensing especially is a very definitive test," said Michael Turner as we pulled into the driveway of the institute's guest house, more of a guest mansion, really. "I should make it clear, by the way, that just because a cosmological constant of point eight gives you a best-fit universe, that doesn't imply that it's the best motivated model, or that I necessarily think it's right. In fact, if you design your theory to fit all the data we have right now, which I did in making this model, then the theory is almost certainly wrong, because it's very unlikely that everything we think we know about the universe at this moment is correct."

Once you admit the possibility of a mix-and-match cosmology, with a little bit of one model and a little bit of another, it becomes much easier to create a real-looking universe. CDM plus a cosmological constant is one way to do it; another is to spice up CDM with elements of a theory that was pronounced dead a decade ago. Joel Primack, the Santa Cruz physicist who had a major role in pushing cold dark matter in the first place, now thinks the most plausible way to build a cosmos is to bring back hot dark matter (HDM) in the form of a neutrino with mass, but not too much. Primack, bearded like his colleague and friend Blumenthal, is shorter and stouter, and is as neat as Blumenthal is rumpled. His graying hair is all perfectly in place, and his beard is close-cropped and tidy. Like Blumenthal, he favors sweaters, but where Blumenthal's are bulky and rough-knit, Primack's are form-fitting and fine. He has a highly developed sociopolitical conscience: for years, Primack spent much of his spare time speaking out against nuclear weapons and Star Wars in favor of arms-control verification, and applying his knowledge of nuclear physics to analyze the technologies involved. Now that nuclear war is less of a threat than it used to be, his focus has shifted to the question of space-based nuclear reactors and whether they're necessary or safe (surprisingly, perhaps, to those who have opposed him on other nuclear issues, he thinks both answers are sometimes yes).

Primack explained his own ideas about mix-and-match universes on the way to lunch at the Santa Cruz student center one hot afternoon. I huffed and puffed, as I tried to take notes, and he breathing easily, negotiating the redwood-covered hills. "Now, I have to warn you that it's only suggestive, and requires a fair amount of speculation," he said. "But it's certainly true that if you added thirty percent hot dark matter to a cold dark matter universe, it gives you more structure on large scales than you'd get for

Joel Primack

PHOTO: ANDREA BERRY

CDM alone, about the right value for large-scale streaming of galaxies, the right cluster-cluster correlations and maybe the galaxy correlations as well.

"Why do neutrinos give extra power? It goes to the original idea of hot versus cold dark matter." The distinction between the two, as Blumenthal had explained, is based on the size of the first structures that could form under gravity. Hot dark matter makes enormous structures first; cold starts out making structures about the size of galaxies. The argument against pure hot dark matter was that it predicts that superclusters, the biggest structures, should be old and galaxies young. "But real superclusters are young," said Primack. "They're still forming. The local supercluster hasn't even begun to fall together yet. It's begun to slow down—we're moving away from the Virgo cluster of galaxies more slowly than we would if all our motion were due to the Hubble flow. But we're still moving away. Galaxies and stars, though, are clearly old. That's why HDM looks nuts."

But what if the neutrino has a mass of ten electron volts, the lower limit of what Tyson's observations allow? "We know quite accurately how many neutrinos there are in the universe," said Primack. "There are about one hundred for every cubic centimeter of volume. So you can easily calculate that a ten-ev neutrino would add up to about thirty percent of the critical density, about a third of what you need to close the universe.

"So now you look at what happens in a universe just at critical density, about five percent baryons, thirty percent neutrinos, and the rest cold dark matter." It turns out that those calculations have already been done: a graduate student named Jon Holtzman, at Santa Cruz in the mid-1980s, had been working on an observational thesis with Sandra Faber. He was going to analyze the first images coming off the Hubble space telescope when it went up in 1986. "He spent two years writing software to analyze the images," said Primack. "Then *Challenger* crashed, and he was finished. He came to me and said, 'I guess I'd better do a theoretical thesis.' What he did was to take a whole raft of theoretical cosmological models—hot dark matter and cold dark matter and cosmological constants and high omegas and low omegas and high Hubble constants and low ones. All sorts of combinations. And he worked out the implications for all of them as far as what you'd expect to see in large-scale structure and in the microwave background—in fact, the COBE people are using Holtzman's thesis to evaluate whatever the satellite finds." (When the space telescope finally went up in 1990, Holtzman took leave from Lowell Observatory in Arizona, where he had gotten a job, so that he could finally use his image analysis software. He ended up spending six months working instead on trying to figure out how to get around the Hubble's faulty mirror.)

"When you look at the effects of CDM plus HDM," said Primack, "you can, to a first approximation, treat them independently. You see what CDM does and what HDM does, and just add the effects together. And what you find, not all that surprisingly, is that HDM leads to structures on very large scales, and CDM leads to them on all scales. So on small scales, you have just the influence of CDM, while on large scales you have the influence of both." You get, in other words, extra structure on large scales, which is the phenomenon cosmologists are trying to explain.

"The unfortunate thing about the HDM plus CDM model, if it's right, is that you're dealing with three different types of matter as major constituents of the universe—baryons, HDM, and CDM. Just having two is weird enough; having three is really weird." (When I got back to Princeton I asked Jerry Ostriker about the hot-plus-cold model; he made a face. "You know," he said, "that's like you're a bad cook. 'Oh, we have too much pepper. Better add some salt. No, too much salt, better add some pepper. I mean, maybe a little basil.' I . . ." His expression became more pained. "If you have two models that don't work, do you make a linear combination of them? Yes, it works better, but I'm not sure it's the way to go.")

Primack would have understood his discomfort, but argued for the theory anyway. "There is one bit of evidence that makes me think you should take it seriously," he said, "and that's the solar neutrino problem." The problem, which astronomers and physicists have been wrestling with for nearly three decades, is that the nuclear reactions at the core of the Sun seem to be generating only between a third and a half the number of neutrinos that nuclear theory says they should. One explanation that was popular for a while was that cold dark matter particles had drifted, over the eons, into the Sun's core, cooling it off and reducing its neutrino output. But other evidence, indirect but compelling, says that the core is just as hot as we've always thought (twenty million degrees or so), and that the answer lies somewhere else.

Another idea is that neutrinos, which come in three varieties (they're called electron, muon, and tau neutrinos) can switch from one type to the other. And if that's true, for reasons that are much too arcane to contemplate, all three varieties of neutrino must have at least a little mass. So far, two independent underground detectors, one in South Dakota and one in Japan, have measured the neutrino deficit; a third, in Russia, was on the verge of issuing its own report as Primack and I talked. The word was that they, too, saw only about a third of the expected number of neutrinos. According to at least one theory of neutrino-switching, that points to a mass for the tau neutrino of about ten electron volts. "If it can really be shown convincingly that the tau neutrino has that kind of mass," said Primack, "the hot-plus-cold model will look awfully good, even if it seems contrived."

CHAPTER EIGHT

NEW SEARCHES

H ow big is the universe? How old is it? What does it look like? What is it made of? How did it begin, and how has it evolved from the beginning until now? Observations like John Huchra's and Margaret Geller's redshift surveys; Tony Tyson's dark matter maps; Sandra Faber's search for large-scale motions; Ed Turner's, Jackie Hewitt's, John Tonry's, and Bob Kirshner's attempts to calibrate the Hubble constant; and the heroically difficult attempts by Lyman Page, Suzanne Staggs, David Wilkinson, and others to detect deviations from perfection in the spectrum and structure of the cosmic microwave background have brought astrophysicists closer to answering these questions. So have intensive theoretical calculations by people like George Blumenthal, Joel Primack, Jerry Ostriker, James Peebles, Michael Turner, and scores of others.

But if the cosmologists are closer to understanding how it all works, no one can say by how much. "It almost fits," David Spergel told me after Tuesday Lunch one day. "I think that we're maybe one step away from figuring it all out." Or the truth could be more as Ed Turner sees it: that existing theories just aren't going to work in their present form. There is a desperate need for more data that theorists can use to constrain and test their speculations. But there are not enough large optical telescopes, not enough microwave-sensitive satellites, not enough large-area charge-coupled devices, not enough hundred-foot-plus radio telescopes, and not enough supercomputers to process the data that are coming in now, let alone the kinds of data astrophysicists would like to have.

The supercomputers will come along in any case; there are plenty of other uses for them besides astrophysics, and their prices will drop as they always have, under the constant demand from other branches of science, from drug companies that need processing power to design new medicines and from aerospace contractors who model complex aerodynamics before they build real planes. What astronomers really need, and despair of ever getting enough of, are photons all up and down the electromagnetic spectrum, for they'll never understand the secrets of the universe without them. It's possible that the present cosmological crisis will recede, but without an exponential increase in the number of photons captured at observatories, another crisis will inevitably come along to replace it.

The situation is not, however, entirely bleak. Now that they know what kinds of structures exist in the universe, astrophysicists are going into the survey business in droves; now that they have hints about how dark matter clusters, they're thinking up new ways to look for the substance that makes up most of the universe. They will soon have the tools to do so. By happy coincidence, two separate teams of astronomers and engineers, separated in space by three thousand miles and in time by only two weeks, in the spring of 1992 took two major steps toward supplying astronomers with the photons they need.

The first group, led by an astronomer named Roger Angel, was working in a cavernous laboratory under the football stadium at the University of Arizona. One day in April, Angel loaded twenty thousand pounds of Pyrex glass into a mold more than twenty feet across and heated it to nearly two thousand degrees. Once the glass was melted, the mold began to spin slowly on an enormous turntable, forcing the liquid to slosh gently toward the edges, its surface forming a near-perfect parabola. Over the next few months, as it kept rotating around the clock, the glass solidified into the world's largest telescope mirror blank, a lightweight disk of glass 254 inches across and an average of only inch thick, its light-focusing concavity already preshaped so the mirror would need a minimum of polishing.

This was the mirror John Huchra had told me about, the one they would use to fix the "holes" in the MMT. Except for another project going on in the middle of the Pacific Ocean, it would have become the centerpiece of the largest optical telescope in the world, a yard and a half wider than the 200-inch on Mount Palomar. But almost fourteen thousand feet above sea level, on the edge of the ancient caldera of the extinct volcano Mauna Kea in Hawaii, another mirror was just being finished that beat Angel's by three and a half meters, or 136 inches—a margin that alone would make a powerful telescope in itself. The ten-meter mirror at the heart of the Keck telescope, built by the private W. M. Keck foundation along with Caltech and the University of California, is even more innovative than Angel's. Like the MMT, its huge light-gathering area is made up of smaller, individual mirrors. Unlike the MMT, there are no holes, and that, according to Jerry Nelson, a Berkeley astronomer who designed the mirror, presented enormous problems.

As the telescope mirror neared completion, Nelson was constantly shuttling back and forth between Hawaii and California, supervising the construction but also talking with the Keck Foundation about plans for a second telescope, just as big, that they had decided to place next to the original one. The Keck telescope had to be renamed Keck I; Keck II will be done in the mid-1990s. "We began thinking about all this at Berkeley in the late seventies," Nelson told me over the telephone between trips. "We felt the community needed a new, large telescope. So a few of us began thinking about how you go about making one. It was obvious that a lot of the processes you go through—casting, polishing, transportation, supporting the mirror in the observatory, minimizing the deformation from heating and cooling—get a lot worse with size. So we thought it might be interesting to look at the idea of making a single mirror out of segments. Radio astronomers have been doing that for years, but of course their wavelengths are large so their tolerances can be relatively large, too. Because the segments stay the same size no matter how big the mirror is, you can, in theory, go as big as you want. We decided ten meters, or twice the diameter of the Hale telescope, was a good, round number."

There were two problems a segmented telescope would have to overcome. One was keeping all the mirrors aligned, which would be much harder with the enormous Keck than with the MMT (Nelson and his collaborators decided in the end to support each mirror segment with three hydraulic arms, each of which could be adjusted to within a millionth of an inch). The second problem was that the Keck's segments, unlike those of the MMT, were not going to be individual, quasi-independent mirrors, but would be mated to form a single, large mirror. A conventional, one-piece mirror is symmetrical; slice it in half, and the curve along the cut edge is always the same no matter where you cut. That makes it possible to grind it automatically: put the blank on the turntable, spin it around, and grind back and forth as it goes.

But while the curve is symmetrical from one side of the mirror to the other, it varies from the edge to the center, growing gentler as it comes in. In a segmented mirror, each segment is therefore not symmetrical: you'd get a different curve depending on how you sliced it. That makes a rotating polisher inappropriate, unless you're clever about it. That's what Nelson decided to be. Each of the Keck's hexagonal mirror segments, six feet across and four inches thick, was deformed, flexed out of shape. Then, still flexed, it was polished in the conventional way. The resulting curve in the mirror's surface was symmetrical, but only until the device that was bending it was removed. The mirror would then relax, and the symmetrical curve in the newly unflexed glass would become asymmetrical in precisely the right way.

"Did we have trouble selling this idea?" asked Nelson. "You bet. Astronomers tend to be scientifically conservative, and they didn't exactly leap at the notion of building a large telescope in any sort of innovative fashion. We had to develop the design, make laboratory demos to show we knew what we were doing. It took four years after we got the tentative go-ahead to convince people we could do it." Now, eight years later, the last of thirty-six hexagonal segments was being lifted into place, and by the end of 1992, Keck I was scheduled to begin operations.

Actually using the telescope would not necessarily be fun. The winter storms that blanket the treeless Mauna Kea moonscape with feet of snow at a time are the least of the problem. Much worse is the air, thinner even than the rarefied atmosphere that leaves scientists at the South Pole gasping. There is a support station at nine thousand feet where astronomers sleep when their shifts are over and where they stay for a day or two to acclimate themselves before heading up to the summit. "The first time I went up there to use the three and a half-meter," Tony Tyson told me (there are several telescopes on the mountain, where the seeing is probably the best in the world), "I went outside to look at the sky. It was covered with haze. I went inside and told the telescope operator and he went out for a look. 'It's crystal clear,' he said. It's the lack of oxygen. When you get old and lose as many brain cells as I have, it clouds your vision." Tyson has taken spectacular images from the summit of Mauna Kea, but he's never seen the sky.

Keck I has more than six times the light-gathering area of the four-meter telescopes at Kitt Peak and Cerro Tololo, but Sandy Faber advised me not to think about size for comparison. "Think about speed," she said. Her point was that the Keck can suck in photons six times faster than either of these quite respectable telescopes. A spectrum of a faint galaxy that would take a four-meter an hour to gather would take the Keck just ten minutes. The time difference is crucial for doing redshift surveys of the most distant galaxies. Plans are already being finalized to mount such a survey from Mauna Kea. Its acronym is DEEP, for deep extragalactic evolution project, and it will probe the universe much farther than even the Century Survey and will, presumably, present astronomers with more surprises. Sandra Faber, who fortunately is on the faculty of a university with a large block of Keck time, is involved. So is David Koo, another Santa Cruz observer and one of the astronomers, along with Richard Kron of the University of Chicago, Richard Ellis of the University of Durham in England and Thomas Broadhurst of the University of London, whose "pencil beam" surveys suggest that Great Walls might be strung out through the universe every one hundred megaparsecs or so.

I saw Koo while at Santa Cruz. He is tall and balding, soft-spoken, and polite. He was rushing to get ready for an observing run at Kitt Peak the next night, but he took an hour or so to talk. "When we did the pencil beams," he said, "we weren't really looking for structure. We were really interested in how galaxies evolve. Instead of doing a survey of a broad patch, we drilled into the sky, looking at an area the size of the moon, about one quarter of a square degree. In photographic plates, you can see perhaps ten or fifteen thousand galaxies on each square degree. We ended up taking spectra of two hundred galaxies in the north, while Ellis and Broadhurst looked in exactly the opposite direction in the south. We had actually been noticing for years that there were larger numbers of galaxies than average at certain redshifts, but it wasn't until we combined our data with Ellis's and Broadhurst's in '87 that the pattern emerged. What you can see in the plots is a periodicity of about 130 megaparsecs (if the Hubble constant is 100). We have subsequently observed in other directions, forty-five degrees from either pole, and we don't see the same periodicity, or at least not to a degree that's statistically compelling. Tantalizing, yes. Intriguing, yes. But not yet compelling.

"In the DEEP survey, we are now going to look out to a level of twenty-fourth magnitude, which is at least two magnitudes farther than before, and take redshifts for thirteen thousand galaxies at that distance. This is where a lot of puzzles are—the nearest of the faint blue galaxies, for example. Keck gives us the ability to look much farther back in time, and in exploring the large-scale structure it's important not just to say what the structure is today but also what it was early on. DEEP will get at such questions as the evolution of galaxies, the evolution of clusters, the evolution of structure, and the geometry of the universe itself."

These are the same questions Bob Kirshner wants to answer, and he is working on a redshift survey covering the southern skies. That's why he was at Las Campanas observatory in Chile working on a medium-size telescope, and why he was available to heckle Tony Tyson at the colloquium at La Serena. "We've tried to be extremely pragmatic," he told me on one of my Cambridge visits, "and figure out a job that a two-and-a-half-meter telescope is the right hammer for. We have prudently stayed on the easy part of the problem, on galaxies that are brighter than the sky itself. We're from the 'good enough is good enough' school. If you can do good, important work and it's easy, hey, why not? Koo and company, on the other hand, are from the 'beat your brains out' school. Their data reduction is agonizing. So, that's where we are. We have data on five thousand galaxies, and although they're sparse they'll give an indication of structure in an area one tenth the size of the visible universe.

"One problem, in the Southern Hemisphere especially, is the lack of a good sky survey, because in order to get redshifts you need to know where galaxies are on the sky. And the problem with making surveys is that these big telescopes were all designed with enormous focal planes to accommodate big photographic plates, twenty inches across. Nowadays, nobody uses plates. So we put detectors this big"—he held up two fingers, pinched almost together—"into a focal plane like this"—he spread his arms out. "CCDs are getting bigger," he said, "but in the meantime Steve Schechtman has come up with a technique that is truly horrifying." You point the telescope at a particular point in the sky and hold it still, letting the sky drift past as the Earth turns. Normally, that would create terrible images, streaks instead of stars and galaxies. But in this case, the astronomers assign one strip of pixels on the CCD chip to a given strip of sky; as the image of a galaxy moves across the chip, so does its assigned row of electron-trapping bins at the same rate.

"We use the survey to mark accurate positions of galaxies on the sky, and then we use that list to run a milling machine." The machine's job is to make holes in a circular aluminum plate, twenty inches across, that correspond exactly to the places where galaxies fall in a given field of view. Then, down at the telescope, the plates are fitted, one after the other, into the focal plane, and the astronomers attach fiber-optic cables at each hole. When the plate is perfectly aligned, the light from one hundred galaxies enters one hundred holes, slides down one hundred fiber-optic cables, and enters a spectrometer. "For historical reasons, this particular spectrometer is called the Tutti-Frutti—its original name was two-dimensional photon counter, and from *two-d* we got to tutti, and added the frutti. So we call the whole setup the Fruit and Fibre. It takes about twenty minutes to change plates, and each exposure lasts for two hours.

"That's the technique. Now, you have to understand that this is hard. There is no book of instructions on how to do this. But what the heck—the Wright Brothers didn't have lessons either. We lay out what we're going to do with incredible care—for us, at least. So in theory each night we can get two hundred redshifts. We've had good luck, and now we've got redshifts for about five thousand galaxies. It takes so long to do this. We started in '86 or '87, and now we have just a couple of strips of the sky. [Actually, they can get up to four hundred per night; each plate is drilled with four hundred holes, and the team can do four runs of one hundred per night—although about 10 percent of the fibers end up not getting any redshifts.] "The way it works is that after a two-hour exposure you go out and unplug the fibers and plug them into another hole. The afternoon before a run, we're out there with colored pens, marking which holes go on which run so we can make the changeover quickly.

"It's interesting—lots of people are building robot manipulators that will move the fibers automatically. It's a good idea, since it's cold out in the dome and it's awkward to switch fibers—you have to reach way over your head. But it turns out that the positioning systems that can move one hundred fibers also take about twenty minutes for a changeover. So far, John Henry is keeping up with the steam drill. The talk from places like Princeton, which is gearing up to do a million galaxy redshifts with an automated survey telescope, is that they're going to be thorough and accurate, like the Coast and Geodetic Survey. But I think it's more fun to be like Lewis and Clark. There's a certain pioneering quality to all this. For example, the fibers are too small to be manipulated easily (they're a third of a millimeter across), so we sheath the ends in hypodermic tubing—the stuff hypodermic needles are made from. But those are still too small, so we embed them in bicycle brake cable. Then the whole bundle of one hundred fibers is fed down to the spectrograph with a length of garden hose. I think Schechtman's wife is still mad at us, since it came from her garden."

Before he's done, Kirshner wants to get several tens of thousands of spectra, covering a single strip of the sky one hundred degrees long by one and a half wide, plus "bricks" to either side of the strip. "Our characteristic depth is thirty thousand kilometers per second. Now, I'm not claiming we've done the statistical analysis. But it *looks* like, my first glimpse tells me, it sure looks like it's clumpy, stringy, has voids and sheets. That's good. We see the structures the same size of what Geller and Huchra saw, but a lot more of them. And we don't see anything bigger. This is the first time the biggest structures contained in a survey are a lot smaller than the size of the survey." Or, as Geller might have put it, they could have seen bigger structures, but they haven't.

"It turns out," Kirshner said, "that there's a theoretical paper by Blumenthal and several others that says, okay, we know how smooth the cosmic microwave background is. Based on that, how big are the biggest structures the universe could contain? The answer seems to be that the biggest voids will be about eight thousand kilometers per second across. If there were anything bigger, our survey could see it, and we don't. What we do see is on the order of eight thousand. This is good. It means maybe after the Great Wall and the Great Attractor, we're coming to the end of the age of greatness and entering the age of mediocrity."

A few hundred miles south, in Princeton, the Coast and Geodetic Survey of redshift projects was also well along in its planning. This survey is so firmly a part of the "beat your brains out" school of observation that the astronomers have recruited physicists from the Fermi National Accelerator Laboratory—Fermilab—the site of the world's most powerful elementary particle accelerator. The Fermilab crew, which includes such astronomy-friendly physicists as David Schramm and Mike Turner, is experienced in the mind-numbing number crunching that is a central part of modern particle physics. That kind of expertise will be crucial in analyzing the terabytes, or trillions of bytes, of data that will flow from the Princeton-Chicago-Fermilab-Institute for Advanced Study-Johns Hopkins Digital Sky Survey (now known as the Sloan Digital Sky Survey, after the Alfred P. Sloan Foundation kicked in eight million dollars, a contribution reported by Jerry Ostriker with great fanfare at Tuesday Lunch).

The chairperson of the project's Science Advisory Committee at the time of that announcement was Neta Bahcall; she's also head of a working group that will study the properties of clusters revealed by the survey. The other five working groups will cover galaxies, stars, quasars, large-scale structure, and serendipity—the last category covers unexpected discoveries, which are sure to be made but impossible to preclassify; Ed Turner is in charge of this one.

"It's really quite impressive," Bahcall told me one afternoon after lunch. "We are planning to get the redshifts of one million galaxies and one hundred thousand quasars—which is on the order of a hundred times as many as we now know of—over a five-year period. It will be a three-dimensional survey that is more or less complete over one quarter of the sky, and will go out to a redshift of about point fifteen, which is about five times deeper than the original CfA survey. We will get redshifts for all the galaxies up to nineteenth magnitude and all the quasars up to twentieth.

"We're going to have a dedicated telescope for this, a two-point-five meter, which will do just this survey. We'll put it at Apache Point, New Mexico, next to the ARC telescope. It is not very large, but it is large enough for this project, and it has an enormous field of view, three degrees across. The light detectors will be thirty CCDs, two thousand by two thousand pixels each, which is unprecedented."

Like Kirshner's survey, this one will be creating its own map of where the galaxies are first or, rather, simultaneously. "The imaging obviously has to keep ahead of the spectroscopy," said Bahcall, "but they will be intermingled. We want to use the twenty percent or so of best seeing for imaging, and as long as we're ahead, we'll do spectra the rest of the time. Our imaging alone is going to produce a map of the northern sky that is something like the Palomar Sky Survey but orders of magnitude better since it goes much deeper. It's done on CCDs, and it is going to be done in four colors, which will give us all sorts of extra information about things like galaxy evolution."

Also like Kirshner's survey, the Sloan will use its maps to direct the drilling of metal plates with holes optical fibers can be plugged into. But it will have six hundred holes per plate compared with Kirshner's four hundred and, more important, six hundred fibers where Kirshner has only one hundred. The fibers will still be attached to the plates by hand ("We considered using robotic arms, but it looked too complicated," said Bahcall). They will be attached in advance, though, the fibers gathered at the spectrograph end into bundles and then into large plugs, scores at a time;

changing plates should therefore take just a few minutes. "Basically," said Bahcall, "however long it takes you to get a spectrum, we can do it six hundred times faster. We're doing this in four years, so it would normally have taken twenty-four centuries. None of this is entirely new technology, of course, but we're pushing the technology to its limit in several areas, and the combination will be very powerful.

"Most of the survey will be done in the spring, when the northern galactic hemisphere is overhead. In the fall, when the southern sky is up, we'll do something entirely different. We're currently planning to do a small area over and over. That let's you go fainter in the imaging, but the main thing is that you'll find a lot of variable objects, including supernovas and variable quasars, and there are all sorts of interesting things you can learn."

Across the lawn from Bahcall's office in Peyton Hall, David Wilkinson and Lyman Page were cooking up their own version of a sky survey. COBE had still not found any fluctuations in the microwave background, and one of two outcomes was possible: either it would find the fluctuations soon, or it wouldn't. In either case, both physicists figured NASA ought to be getting another cosmic background satellite in the pipeline to keep looking (desperately, by this point) for the fluctuations that simply must be there, or to take a harder and more detailed look at what fluctuations COBE had finally discovered. "Some of us began thinking about this a year ago," Wilkinson told me one afternoon, about a year after I had gone to Green Bank. "There comes a time when the science is compelling, and the technology is sweet, and you know it won't get much better, and you have to start pushing. The detectors on COBE are twenty years old. I think the time is right."

I asked him why a new satellite, when the South Pole and balloon-borne observations kept getting more sensitive. "Well, you still want to get above the atmosphere, and you really want to get away from the Earth altogether—earthshine and the radiation belts are problems you just don't want to deal with. I'm proposing putting a satellite out at L2, one of the libration points between the Earth and the Sun." A libration point is a place where the gravity of two celestial objects just balances, where an object like a satellite will be pulled equally in two directions and so will stay put with a minimum of adjustment by positioning rockets. "L2 is about a million kilometers away. From there, the Earth and the Sun both look about a half a degree wide. You get a lot of really nice, clean viewing angle. NASA isn't taking any proposals for this class of satellite right now, but if you wait until they do, you're going to be too late. We've really got to get the community behind this now, so we're ready when the time comes."

What a new microwave satellite won't find, except perhaps indirectly, is an answer to the question of what the universe is really made of. The dark matter accounts for most of the mass that exists, but even Tony Tyson's dark matter maps can only say where the stuff is located, and how much of it there is, not what it is (except, again, indirectly: the way dark matter clumps together can rule out some candidates). All of the searches discussed so far are ways of trying to get at the dark matter in the aggregate, to uncover its properties in clumps that weigh as much as hundreds of billions or even hundreds of trillions of Suns. But if, as most astrophysicists still suspect, the dark matter is made of exotic elementary particles, then the answer to what the universe is made of can be found, in principle, by detecting a single subatomic particle. If the Soviet-American Gallium Experiment (SAGE) or the Homestake Gold Mine detector or the Kamiokande detector can show that neutrinos have a tiny bit of mass, the search is over. Or if particle physicists can show convincingly that the still-mythical axion exists, or some other weakly interacting massive particle (WIMP) that devolves from the equations of grand or super unification, the dark matter problem will be solved.

That is in essence what the Center for Particle Astrophysics (CPA) at Berkeley is all about. The center was set up by the National Science Foundation, one of twenty-five science and technology centers funded by the foundation beginning in 1989 to attack basic research problems from several angles at once. The

Bernard Sadoulet   PHOTO: BENJAMIN AILES

CPA's job is to understand the nature of dark matter through both macroscopic searches (the DEEP survey is one of the center's projects) and microscopic searches. When David Schramm waylaid me, I was on my way to see Bernard Sadoulet, a professor of physics at Berkeley and the director of the Center for Particle Astrophysics. The center has projects going all over the world, but its offices are shoehorned into a handful of rooms in the physics building on the Berkeley campus. The center's library is a room perhaps twelve feet wide by fifteen long, with current scientific journals, a couple of computers, and not much else; it also serves as the waiting room for the director's office.

Bernard Sadoulet is a bearded Frenchman, his accent undiluted by years of working in the United States. His office was dominated by a poster-size color photograph showing roughly circular blobs of light on a dark background, yellow, red, and blue; some of the blue blobs were stretched out into arcs. "Ah, yes," he said. "Tony Tyson was generous enough to send me that. His work is very beautiful and very important." Sadoulet himself was trained as a particle physicist, and he had been scrambling to get up to speed on astrophysics since he signed on as director three years before. He urged me to speak with the astrophysicists about their part of the project (hence my conversation with David Koo), but explained the particle search himself. "All of us would agree," he said, "that if omega is bigger than about point two, you need some form of nonbaryonic dark matter. As Joel Primack likes to say, 'This is the ultimate Copernican revolution.' Copernicus told us the Earth doesn't revolve around the Sun, that we're not at the center of the universe. Now we seem to be learning that we're not even made of what the rest of the universe is.

"If you say the dark matter is nonbaryonic, then you have many possibilities for what it could be. There are neutrinos, of course, and you'll remember that ten or fifteen years ago, there was big excitement when a Russian group claimed it had discovered massive neutrinos. This result was not correct, but now there is excitement again, because of the SAGE experiment. But it looks now as though the neutrinos will not have enough mass to close the universe, and in any case, neutrinos alone have a problem because they do not make structures smaller than superclusters."

"So that's one case. Another is that there are much heavier particles, with a mass of about one billion electron volts, or GeV— the so-called WIMPs that make up what we call cold dark matter. If these particles have a GeV of mass, you can compute how many of them are in the halo of our galaxy. It turns out that there are so many that there should be billions passing through a human body every second. Sometimes, very rarely, one of these WIMPs will interact with baryonic matter; on average, if you have a kilogram or so of some target material, there will be one event every ten days, and our job is to detect it."

Sadoulet explained that when the interaction happens, the incoming WIMP, slamming into an atom in the target, transfers its energy of motion to the atom. If the atom is in the human body, nothing happens: the energy is far too small to have any consequences. But if the atom is in a crystal of, say, germanium 73, two different things happen. First, the impact knocks some electrons loose from the atom; the atom is ionized. Second, the atomic nucleus is knocked from its position. It isn't knocked loose: it's more like the way a bedspring would react if you dropped a baseball on it. The atom recoils, then returns to its position, but in doing so it sends a wave through the rest of the atoms it's attached to. This wave moving through the crystal is known as a phonon.

"Both of these phenomena are, in principle, detectable," said Sadoulet, "and, since we're looking for a needle in a haystack, we want as much information as we can on each event. So we do two things. We put a voltage across the crystal so we can measure the very slight increase in current from the freed electrons. And we also put a thermistor on its surface." A thermistor is a device that conducts electricity but whose resistance changes with its temperature. When a phonon ringing through the crystal hits the thermistor, it heats it up just a hair, and the resulting change in resistance is measurable—it's a clue as to how much energy the phonon carried and thus how much momentum the incoming WIMP was packing. "So you're measuring two things at once," said Sadoulet. "You get the amount of ionization caused by an interaction and the strength of the phonon. By measuring two different quantities, you have redundancy—if you get ionization but no phonon, for example, you have to be suspicious. You also can measure the ratio between the two quantities, which should follow a characteristic pattern."

In order not to be swamped by spurious phonons arising within the detector, the whole thing is cooled to twenty or thirty thousandths of a degree Kelvin. And to avoid impacts from other, non-WIMPy particles, the whole apparatus, about the size of Suzanne Stagg's microwave detector, is going underground, where several yards of dirt and rock will screen out cosmic rays, a form of subatomic junk that flies in from space. The first test run will take place thirty-five feet down in a chamber that's been excavated across San Francisco Bay at the Stanford Linear Accelerator Center. "We see one to two years underground at Stanford," said Sadoulet, "and then a move very far underground, probably to the Oroville Dam, where we've had some experiments in the past. The official plan is to start our operation in the fall, although that could slip."

Finally, and just to make sure they cover all the possible bases, scientists at the center are mounting a search for the form of dark matter long since discounted by most theorists (Jerry Ostriker and Jim Peebles at Princeton are two significant exceptions): ordinary, baryonic matter in the form of giant, Jupiter-size planets; small, dim stars (brown dwarfs) or even black holes lazily orbiting around galaxies like the Milky Way to form an invisible halo of gravity-generating mass. While Sadoulet and his crew are looking for WIMPs, a young physicist named Kim Griest, also on the staff of the CPA, is part of a large group hunting for (no kidding) MACHOs.

"It stands for massive compact halo object," said Griest, whose office is around the corner from Sadoulet's. "I came up with the acronym, and I freely admit that it plays off WIMPs. When I came up with this acronym, it was so high on the stupidity scale that I said why not?" Griest, like Sadoulet, considers himself a particle physicist, but he's perfectly happy with a detour into observational astrophysics. "Dark matter is my thing," he said. "I'm trained as a dark matter theorist, and I'm interested in the dark matter whatever it's made of. I happen to think that exotic particles are the best guess as to what it is. But looking for particle dark matter is very hard. The second best idea is that dark matter is baryonic. Despite what some people say, I think you can explain just about all the dynamic evidence—the rotation curves of galaxies and the extra mass in clusters—with baryonic matter, and it's certainly clear that at least some of the dark matter is baryons.

"So what is it? It could be brown dwarfs, or Jupiters, or black holes, stuff that clumps on scales of between the mass of the Earth and about one hundred solar masses. The technique we're using to look for them was first proposed by Bohdan Paczynski at Princeton, in 1986. At the time, most observers didn't pay attention because it called for observing a million stars a night for years. But technology advances, and maybe because we have lots of particle physicists in our group we're not afraid of big collaborations the way traditional astronomers are. This project, for example, has twelve people. I'll bet that Tyson goes to Chile with, what, one other person? But we have to reduce enormous amounts of data— sixty-four megabytes every five minutes—so we need software engineering and lots of stuff. The Princeton million-galaxy redshift survey is comparable. In 1986, the kinds of computers and CCDs we needed to do this project were just too expensive, but now there's really nothing in our way."

Griest and his collaborators are looking for MACHOs with the same technique Ed Turner, Jackie Hewitt, and Tony Tyson use: gravitational lensing. Galaxies lens quasars and they lens other, more distant galaxies, but if MACHOs exist, they will occasionally pass in front of background stars in the Large Magellanic Cloud. When they do, the stars will be magnified; they'll temporarily look brighter, by between 30 and 500 percent, and then as the MACHO passes by will fall back to their original luminosity.

"Unlike quasars," said Griest, "which we don't even know what they are for sure, and which flicker intrinsically, stars are extremely well understood. So if we see an event, we will be able to say whether it's real or not. There are a lot of different kinds of signatures you can expect in the light curve of a lensed star, depending on how directly the MACHO passes in front of it and also on how massive the MACHO is." If it has the mass of the Earth, the event will last an hour. If it's like Jupiter, it should last about five days. If it's got the mass of the Sun, it should last a few months. For a hundred-solar-mass black hole, the event should last a hundred years, so the observers would only see a piece of the light curve. Griest told me that MACHOs can't be much bigger than that, presenting an argument I'd first heard from Jerry Ostriker. Otherwise their gravitational disruptiveness as they plunged through the disk of the Milky Way would long since have distorted its shape.

The first step is to identify three million or so stars in the Large Magellanic Cloud that are not intrinsically variable. "The way we do it," Griest said, "is by taking pictures of the Large Magellanic cloud and measuring the brightness of all the stars in every field. We'll then return to the same fields over and over to see if any stars have varied. If so, we weed them out. Once we've identified our candidates, we'll monitor them for a year. If we see nothing in that time, we'll be able to say for sure that there's no significant baryonic dark matter in the halo, which is the result I expect. (A similar search was getting under way at the same time, run by Paczynski and several colleagues in Poland.)

"If we do see something, we'll want a minimum of ten gold-plated events a year, enough to convince other astronomers that this is real. Jerry Ostriker has criticized us, saying, 'Oh, you'll be confused by some sort of weird variable star.' But that's why we're taking data in two colors—gravity is color-blind, so a lensing event should be seen in all colors. Variable stars tend to brighten in one band more than another. Besides, we'll be monitoring several different types of stars. If we see real lensing events, we should see them with all types of stars, and it's almost inconceivable that we'd find a new, weird kind of variability that is identical across star types. Also, you should get more grazing events than full lensing events, in a predictable ratio, each with its own brightness signature. Two years ago, when we first started, we were skeptical ourselves, but we have brought the project along, and now the science is solid. The only variable here is our competence."

The MACHO men and women might have liked to base their search in Chile, where the Large Magellanic Cloud swings high overhead and where the skies are usually clear and dry. Unfortunately, there were no spare telescopes lying around unused, and the constant monitoring required by the project called for an instrument fully dedicated to the task at hand. The group settled for Australia. "We've been given exclusive use of the fifty-inch telescope at Mount Stromlo, in Australia," said Griest. "Unfortunately, it rains a lot there. But if we wanted to go to Chile, we would have had to build our own telescope, which would have been expensive and would have set us back a year or two. It's definitely a good, exciting time to be a cosmologist. There's a lot of upheaval."

CHAPTER NINE

"IT'S LIKE SEEING GOD"

I regard cosmology as pretheoretic," Joel Primack told me on my visit to Santa Cruz, "the same way geology was before plate tectonics, or physics was before Newton—just a collection of facts. That isn't to say Newton provided the final theory of physics—just the first. When we finally understand the initial conditions, we'll be able to construct our theories based on data rather than on sand." Throughout the year and a half I spent traveling with and talking to cosmologists, it became clear that virtually all of them would agree with Primack. A half century after Hubble proved the existence of galaxies beyond the Milky Way and showed that the universe is expanding and a quarter century after the discovery of the cosmic microwave background radiation and the firm establishment of the Big Bang, astronomers are still constructing their theories based on the tiniest scraps of data. It's no wonder that cosmology has entered a period of crisis. As Michael Turner said, "If you design your theory to fit all the data we have right now, then the theory is almost certainly wrong because it's very unlikely that everything we think we know about the universe at this moment is correct."

The hunger for data is universal. "Give the observers time," John Huchra said.

"We really need a national twenty-five-meter telescope," Tony Tyson told me.

"We just need more observations," said Ed Turner.

And the data kept coming in, albeit slowly. Most of the time, astronomical observation is methodical work; even when you make a great discovery, it often takes months of computer analysis to realize it, or even years, the time it took Huchra and Geller to accumulate the redshifts that led to the great voids and Great Wall. During none of my visits to observatories or university offices did anyone ever jump up and say, "My God, look what I've found!"

But if I'd skipped the trip to Cerro Tololo and instead tagged along with Tony Tyson on his next run at Kitt Peak, just two months later, I might have seen something. Shortly after that five-night run ended, Tyson gave me a call. "We've got something very interesting here," he said. "If you promise to keep it quiet, I'd like you to come on over and take a look." He was in his office, at Bell Labs' Murray Hill campus, about forty-five minutes north of Princeton. I had been there before, and the routine at the entrance was familiar: I had to fill out a form stating my name, affiliation, and purpose, and then the guard called Tyson, who had to come downstairs and escort me up. While I waited, I looked at posters warning me to shred my documents rather than throw them into Dumpsters where competitors could crawl in and retrieve them. As Tyson had pointed out in earlier conversations, Bell Labs does leading-edge scientific research, but it is primarily an industrial-design lab, and the competition with other labs to design profitable products is fierce.

What wasn't at all familiar was Tyson's office: he had moved into a new one since the last time I had been there. "I keep moving to avoid my creditors," he said, but even he couldn't bring himself to fool around any longer. "Okay, here it is," he said. "I've got to say that this latest discovery is galvanizing. I don't really like to publish before I've done a complete set of careful observations, but this one is really too important to wait on." He and Gary Bernstein, who was with him for the run, had finally made an observation that he had been planning for some time, and which Ed Turner had, for his own reasons, been urging as well.

Tyson had trained the four-meter on Kitt Peak, the twin to the big telescope at Cerro Tololo, on 0957, the original double quasar that was the first gravitational lens ever seen. Ed Turner was interested in knowing how much mass the lensing cluster of galaxies really contains and how it is distributed, to help settle the question of how the time delay between flickers of the two quasar images translated into a precise distance to the lensing cluster of galaxies and thus into a hard number for the Hubble constant. Turner knew that existing estimates of the mass distribution were seriously iffy, and that Tyson's arc-finding technique had a good chance of nailing it down.

"We didn't expect to see anything unusual. We were using this field for sky calibration," said Tyson. The reason is that the foreground cluster is a relatively sparse one, and that while there was clearly enough mass in it to split a quasar in two, there probably wasn't enough to stretch the faint blue background galaxies into appreciable arcs. "What we did see blasted our socks off," said Tyson. "There was evidence for arcs in the background galaxies, which means there is a concentration of mass there that is far greater than anyone has suspected. But there isn't any significant cluster of galaxies there—just two or three, one of which is evidently the lens itself.

"The nice thing about arcs is that they let you have a precise answer for the distribution of mass in the foreground, which is the least well-known aspect of the 0957 system, the one that leads to most of the uncertainty in the Hubble constant." By plugging in his new data on the distribution of mass in the foreground and using Press's and Hewitt's estimates of the time delay, Tyson had done a back-of-the-envelope calculation to come up with a better estimate of the Hubble constant. Until now, the Hubble constant, as calculated from the 0957 system by Bill Press and Jackie Hewitt, had been the most reassuring of recent measurements of that crucial number: unlike virtually everyone else, who were getting numbers up around eighty, their figure was low, in the neighborhood of forty, implying a universe old enough to account for the ages of the stars. "We've redone the calculations based on our mass estimate for the system," said Tyson. "We get a figure of between sixty-eight and eighty-seven.

"Ultimately, though, this may be just as important for being the first example of a big pile of dark matter without a significant amount of visible matter associated with it. Although there are just two or three galaxies visible in the foreground, they contain as much mass as you'd normally find in a good-sized cluster of dozens of galaxies. This is an amazing discovery. You've got a handful of galaxies with the mass of an entire cluster. There are piles of this dark matter stuff where there is not much visible matter." The idea that the advocates of CDM had come up with, almost in desperation, to make the galaxies in their theory match the arrangement of galaxies in the real universe—the idea of biased galaxy formation—was supported for the first time by observation. The idea that somehow, somewhere, the universe is hiding ten times as much matter as anyone has ever been able to see directly or indirectly—an idea based on probability arguments and on the predictions of the unconfirmed theory of inflation—now had evidence, if only from a single observation, to back it up.

There was another a year later. At the January 1992 meeting of the American Astronomical Society, in Phoenix, a team of astronomers announced that the ROSAT X-ray satellite had detected an enormous cloud of hot gas floating within an unusually sparse cluster of galaxies. The gas molecules are whipping around so fast that the cloud's own gravity is nowhere near enough to hold it together. The only plausible explanation is that there are some twenty trillion Suns' worth of dark matter in the cluster, perhaps thirty times as much dark matter as there is gas and stars. Like the 0957 system, this one seems to show that dark matter is hiding in unexpected places.

Another deep-space image with direct bearing on the cosmological crisis came from the Hubble space telescope—a surprising fact, perhaps, to nonastrophysicists who have watched the Hubble's public relations image swing from prelaunch hype that promised sharp views of everything from Jupiter to just short of God to postlaunch press, which implied the telescope was barely a step up from space junk. Astronomers have known ever since the satellite's mirror troubles showed up that the Hubble was hobbled but not crippled. The essential problem is that the mirror wastes most of the light falling on it; only about 15 percent of the light from a given star or galaxy ends up concentrated and focused. The resulting images are about as sharp as they would otherwise be, just dimmer. This makes the Hubble pretty useless for looking at very dim, diffuse objects like distant galaxies but not so bad at taking pictures of bright or concentrated objects—the planets, nearby galaxies, stars, quasars. A few months after it was launched, the Hubble discovered a gigantic storm system on Saturn and took the first photos ever that showed Pluto and its moon, Charon, as distinct points of light. It has resolved a glowing ring of gas surrounding Supernova 1987A, in the Large Magellanic Cloud; and it has come up with evidence for a black hole at the core of the galaxy M87, in the Virgo cluster (the evidence is in the form of a sharp increase in brightness toward the core, implying a tight knot of stars huddled around some enormously massive but tiny object).

All of these discoveries came as part of the telescope's official observing program, but the single most cosmologically significant picture the telescope took in its first two years of operation did not. It came instead from the Snapshot Survey, a scheme dreamed up by John Bahcall to take advantage of the Hubble's occasional idle periods. Because the space telescope is in a relatively low orbit, the Earth fills much of its "sky." Because the sensitive CCDs on board would be fried by looking at such a bright object—and by looking at the Sun or the Moon as well—the Hubble sometimes has to sit and wait for a few minutes before the next celestial object on its complicated schedule comes over the horizon. While it's just sitting there, John Bahcall and other astronomers reasoned that the Hubble might as well take some quick shots of things it happens to be pointing toward, just in case something interesting turns up, like quasars, for instance. Because the Hubble's forte is sharp resolution, the survey team, which included John and Neta Bahcall, Don Schneider, and postdoctoral fellow Dani Maoz, Tuesday Lunchers all (Schneider is a frequent guest host for John Bahcall when neither he nor Jerry Ostriker can make it) reasoned that quasars double- or triple-imaged by a gravitational lens, but too narrowly separated to show up from Earth, might show up in the Snapshot Survey.

The team made a list of 354 quasars out of the 4,000 or so now known that have especially high redshifts (increasing the chance that a galaxy will lie in front of them, cosmological constant or no) and are intrinsically bright (making second and third images easier to see). They then figured out when each would be close to the telescope's line of sight in between formal observations and programmed the Hubble to start taking snapshots. If the matter/energy density of the universe is dominated by a cosmological constant, the universe is older and bigger than a straightforward interpretation of redshifts implies; more galaxies will lie on lines of sight to quasars than would without the constant, and the survey should have turned up at least a handful. It found only one, a quasar called 1208 + 101 whose double images are separated by only a half arcsecond. (It would have found four or five, except that several quasars that fit the selection criteria are already known to be lensed and were thus deliberately left out.)

The most straightforward explanation for the low turnout is that the best fit universe of Michael Turner doesn't fit. If there is a cosmological constant, it is almost certainly not big enough to close the gap between the amount of matter astronomers can observe, directly and indirectly, and the amount it would take to flatten the universe, to bring omega up to one and guarantee that the cosmic expansion will forever slow but never stop. (The deflation of the cosmological constant idea is consistent with the blob of dark matter Tyson seems to have found in the 0957 gravitational lens. A cosmological constant could close the universe but it would be smoothly distributed through space not gathered into piles.)

Turner himself was not particularly disturbed by this revelation. "In my heart of hearts," he told me, driving to Princeton from the Philadelphia airport, "I think the cosmological constant is zero." What he was waiting for much more anxiously was some sort of breakthrough on the cosmic microwave background. "Most people look at the CMB as the cleanest test of the theory," he said. "The redshift surveys are really kind of impressionistic. What they're trying to do is compare the universe we see to the universe you'd predict from the theory. You want to know if the simulations and the real universe come from the same machine. Suppose I gave you a sequence of the two letters h and t: htttthhtttthththhhtththhtht. Then I asked, could this sequence have been generated by a coin flip? The simple-minded answer is that you expect the number of heads and tails to be equal. But that's only on average. If you flipped a hundred coins, you'd usually expect to see something close to fifty-fifty, but not always, and you'd rarely expect to see a split of exactly fifty-fifty. So does ten heads in a row mean it wasn't generated by coin flips? No. Does a hundred? Probably.

"Now, suppose instead of giving the sequence to you on paper, I read it to you. And suppose I'm mumbling; then you'll hear some of the letters wrong. In astronomy, measurements are not perfect, and besides, we're measuring the distribution of light while the simulations are showing the distribution of matter. And you want to measure the distribution of matter within a certain volume, but our observations are flux limited. You see that in the CfA pie wedges. Things crap out at the edge because you're seeing only the brightest galaxies at high redshift. Then it gets worse. With a coin, you need to know it's flipped properly—you don't want to miss any flips, and you don't want to count any that end up in a crack or get caught by the flipper. The catalog of galaxies Margaret and John used for the CfA survey is based on photographic plates, which are not very accurate. So while the CfA survey is very interesting, and showed more structure on large scales than people expected, it's not completely convincing, although Margaret Geller is so positive about it that she has me half convinced anyway."

The microwave background is considered a cleaner test of theory than a redshift survey, because it's not based on any sort of selection by the observers: COBE and the other microwave-detection experiments aren't choosing objects from a catalog and measuring them but are instead measuring, indiscriminately, every photon of microwave light they can see at around 2.7 degrees Kelvin. Second, they're not forced to match up the structure they see in visible matter with the structure they've deduced for dark matter, the way most surveys are forced to operate. Third, the microwave background comes from a time in the universe's history where the initial fluctuations, presuming they were really there, hadn't been distorted out of recognition by the "messy astrophysical processes" Ed Turner talks about. The largest-scale fluctuations, which COBE was looking for, would not have had the time to be processed at all by the time of decoupling; they would be truly primordial, imprinted on the universe early on in the first second of its existence and unchanged after that.

At the time Mike Turner and I talked, COBE had been orbiting the Earth for more than two years; its handlers had reported after the first year or so that the spectrum of the cosmic microwaves, the relative strength of the signal at different wavelengths, hugged the ideal theoretical version of a black-body curve nearly perfectly. But on the question of fluctuations, hot spots in the microwaves that would finally give a clue about the origin of structure in the universe, there was nothing. The cosmos that COBE saw with its microwave-sensitive eyes was still utterly smooth. But it was clear to Turner that something would have to happen soon.

"The observations by COBE and also by the ground-based and balloon-based experiments are now at the level of sensitivity where they should be starting to detect something if the theories are right. There are about ten groups, and four or five of them have detected some anisotropy. But making the claim publicly is going to be tough. Everybody's going to be gunning for the first guy who steps out. You have to show that the signal you see is not the instrument and not the galaxy and not the Moon and not the atmosphere. Everyone knows how good Dave Wilkinson is. He discovered anisotropy ten years ago, then he found out he had made a mistake; it was a signal from the galaxy. But this could break any day now."

By late March, rumors were beginning to circulate through the astrophysical grapevine that the break would come at the April meeting of the American Physical Society in Washington. The COBE team would have something important to say. But none of them would say what it was, not to the press and, uncharacteristically, not even in confidence to other astronomers and physicists. David Wilkinson showed up at Tuesday Lunch and wouldn't reveal a thing. "If you can't make it to Washington on the twenty-fifth," he said, "come over to Jadwin at two-thirty in the afternoon, and we'll have our own presentation here." Wilkinson and George Smoot, the Berkeley physicist who would be making the COBE team's announcement in Washington, showed up at a National Science Foundation–sponsored conference on cosmology at the University of California, Irvine, in late March and wouldn't say anything. "Frankly," said Neta Bahcall after she returned, "it was kind of irritating. A lot of top cosmologists were at the conference, and everybody knew that they had an important result and they wouldn't talk about it. It seemed like a real waste."

Whatever the result was, it would be dramatic. Either COBE had found structure in the microwave background, presumably supporting some theories and rejecting others and bringing the crisis closer to a resolution, or it had found no structure at a new, exquisite level of sensitivity, eliminating all theories and perhaps even throwing the Big Bang model itself into jeopardy. I asked Lyman Page what he had heard and what he thought. "Be there," he told me. "That's all I can say."

He would be there; the session of short talks during which Smoot would make his announcement would cover all the latest results, not just COBE, and Page had been asked at the last minute to report on what the balloons were saying. Unfortunately, the session was starting at eight in the morning, and the data reduction wouldn't be done until late the night before. He didn't want to head for Washington until it was. He ended up taking a train that left Trenton at four forty-nine A.M. and got to Washington at seven fifty-five A.M.

"It was pretty wild," he told me later. "I talked to Dick Bond in Toronto at nine the night before—he's helping us analyze our data. He said, 'I don't have it yet, but call me before you leave.' I got up at three and took a shower and got dressed. I was just about to call him when the phone rang—it was twenty-one minutes to four. He said, 'Okay, here are the limits.' I had saved this little space at the bottom of my viewgraph, and I wrote it in and headed for the train. I rode down with Itzhak Sharon, who was giving a talk on gamma rays from uranium. He tried explaining it to me, but I have to admit I didn't follow it very closely. I had a lot on my mind, and I ended up getting two hours of sleep on the train, which I needed. I got back that night and there were millions of electronic mail messages saying stuff like 'Lyman, you looked kind of tired up there.'" Page's group had what Dick Bond calls a "soft detection." There was probably a signal there, a significant change from what they were able to say a year earlier but not a strong enough signal for them to go public.

COBE's detection, though, if there was a detection, had to be strong or the astronomers and physicists who had traveled to Washington and dragged themselves out of bed for the announcement might turn ugly. Alan Guth and Tony Tyson were there; so was David Spergel, who had driven down that morning with six of his students to be present for what would likely be a historic announcement. The meeting room at the Ramada Renaissance Techworld Inn where the session was scheduled was deemed too small so it was moved to a medium-sized auditorium, which filled up quickly as eight o'clock approached. I learned later on that the meeting room at Princeton, which is twice as big, was filled as well with people who didn't have the time to travel (and who probably thought that hearing the news from David Wilkinson was more appropriate, and more than adequate).

At precisely eight, George Smoot stepped to the front of the room, a slim man with a beard, glasses, and a smile that he couldn't control. Ed Turner had predicted he would start by showing a slide with an artist's rendering of COBE, and then go into an excruciating exposition on the satellite's engineering details. "It's a subtle form of torture that experimentalists like to inflict," he told me. "It's a way of saying I know the answer and you don't."

But Smoot went right into his presentation. "We're here to talk about the first year of COBE DMR data [the differential microwave radiometer is the instrument that looks for anisotropy]. If we don't see any fluctuations, then something other than gravity is responsible for the structure of the universe. We worked hard to get here. Well, here's what the microwave sky looks like." He had the projectionist flash the classic image showing an oval all-sky map, bland pink except for a reddish spot near one edge and a bluish spot directly opposite. "Fifteen years ago, at this very meeting, we presented evidence for a dipole in the microwave background." Smoot was the presenter, having flown a microwave radiometer on a high-altitude plane and measured the first microwave fluctuations ever seen. The dipole, as Dave Wilkinson had explained to me in Green Bank eleven months earlier, is a simultaneous redshifting and blueshifting of the microwave background in opposite directions on the sky caused by the Earth's motion through the universe. "You can remove this from the data," said Smoot. The next slide showed a microwave sky with no features at all. "And the next thing you want to see is the quadrupole. This is pretty special. We've been looking for it for a long time. Okay, so what else. You can also look for the octupole, the hexadecipole, and on up. Fortunately, DMR is equipped to deal with these."

These higher-order terms are predicted in all theories that attempt to build structure in the universe, by means of gravity, out of primordial fluctuations. That's because any process that creates fluctuations in the underlying fabric of space should create them at all sizes, just as the wind acting on the ocean creates ripples, waves, and large swells all at the same time. Inflation, in particular, makes the specific prediction that the amplitude of these fluctuations—the height of the waves—should be about the same at all scales. COBE was sampling all scales from ninety degrees, which covers a quarter of the universe at a stroke, down to about ten, which covers a three-hundredth, so it could, if it found any fluctuations at all, show whether inflation was still a plausible theory.

"Okay," said Smoot. "The bottom line is that we've detected an

George Smoot at the ceremonial South Pole (the actual pole is several yards away).

COURTESY OF LAWRENCE BERKELEY LABORATORY, UNIVERSITY OF CALIFORNIA

anisotropy in the quadrupole term at a level of seventeen microkelvins, plus or minus five, or one part in five times ten to the minus sixth. We've also detected anisotropy in all the higher-order terms. You can compare this with the scale-invariant power spectrum of fluctuations like the one you'd expect for inflation. This is what you'd see." He flashed a slide with a chart showing a solid line, dropping from high on the left side and then hovering around a horizontal line going all the way across the chart. Along it were the measurements of the DMR—not individual points, which would have represented perfect measurements—but points at the centers of short vertical lines, the so-called error bars. The true values of the measurements lay somewhere along the vertical lines. The lines themselves, though, hovered right around the theoretical line predicted by inflation.

"As you can see," said Smoot, "it's a pretty good fit. We don't go to a small enough scale to see fluctuations that represent any structures we can see in the modern universe, like superclusters of galaxies, but at the large scale this plot is consistent with inflation." In the back of the room, Alan Guth was beaming.

After that, the rest of the program, Page included, was somewhat anticlimactic. Smoot had dropped a bombshell, as implicitly promised. The American Physical Society press office realized it and arranged for him to drop it again at a press conference. He cheerfully agreed, and at noon he and three other COBE scientists were ready, sitting at the front of a cavernous exhibition hall in the hotel basement, elevated on risers and sweating under the unnaturally bright lights of hastily assembled TV crews. Once again, Smoot spoke first. "We now have direct evidence of the birth of the universe and its evolution," he said. "The English language doesn't have enough superlatives to convey what I have to tell you. This is a very emotional event for me. We have observed the largest, most ancient structures in the universe, fifteen-billion-year-old fossils created in the very early universe. Until now, there was a conflict between the early universe, which is very smooth, and the modern universe, which is very structured. Twenty-eight years ago, the discovery of the cosmic microwave background radiation vaulted the Big Bang into the forefront of astronomical theory. Now, with this, the first cosmological satellite, we are probing into the Big Bang. People started looking for these structures soon after the microwave background was discovered, but only now have we found it.

"We've taken the first year of data, from December of 1989 to December of 1990, and fitted it together like a giant jigsaw puzzle of the early universe. It's a cosmological survey of the whole universe, a map. You know how it looks when you take a globe of the earth and flatten it into a two-dimensional map. We've done that with the universe. Our own galaxy is the most obvious structure in the map, but when you remove the galaxy, the structure remains.

"The temperature variations we measure are extremely small, only about thirty-one millionths of a degree Kelvin—a tremendously difficult measurement. It's like trying to find variations of an inch on top of Mount Everest, or imagine you want to measure the distance from New York to Chicago using your car. You set your odometer and drive, and when you get there you look and read off the mileage; it's accurate to within a mile or two. What we've done is get in the car and drive, but we've kept track of every time we changed lanes or pulled in for gas, and calculated the extra distance that comes from the road not being a straight line, and we had to be accurate to within a foot.

"And what we've seen is very revealing. The universe is full of structure at all sizes—we can see structure as small as five hundred million light-years and as big as ten billion. This satellite is a time machine—it sees ripples in spacetime laid down earlier than the first billionth of a second. If you're religious, it's like seeing God. These ripples are primordial—too big to have evolved between the Big Bang and the time we see them. They are older than the stars, older than the galaxies, even older than the microwave background. They're bigger than anything you can see with optical telescopes. The smallest is bigger than the Great Wall of galaxies."

This caused a little confusion among the reporters. What Smoot meant was that the warm spots have grown with the expansion of the universe so that if you could see them *today* they'd be bigger than the Great Wall. The entire visible universe would have fit comfortably between the Milky Way and Andromeda at the time the microwave background was generated, and the structures seen in COBE's sensors were correspondingly smaller.

"Until now, we've really had only a handful of evidence about the Big Bang, and now we suddenly have a lot more. The simple inflationary Big Bang model is one that predicted these structures. But it's got to be checked very carefully. One way is through additional COBE measurements, which are still going on. Another is through measurements by other groups, and these experiments are going on too. I predict a gold rush to make these observations."

Smoot sat down, and Ned Wright of UCLA took the microphone. "The fluctuations we're talking about today appeared during inflation, but long before the structures we see today. The hot and cold spots on the map are like hills and valleys in spacetime. Matter responds to the valleys and falls into them. The big question in cosmology is: how do you go from a very smooth universe to a lumpy one? The simple answer is: gravity. So now that we've seen this structure with COBE, we know how deep the valleys are, and we can ask, is the amplitude consistent with modern structure? Yes, with a condition: you need dark matter to help make the structure." A question from the floor: if you surveyed a big enough area to take in the modern-day descendants of these hot spots, what would they look like? "They'd be enormous volumes of space with a slightly higher density of galaxies than average."

What you don't need, evidently, is cosmic texture; the microwave hot spots Neil Turok and David Spergel had said COBE should see if their model was correct were nowhere to be seen in the COBE maps. I looked for Spergel in the hall after the press conference and asked him about it. "Oh, we're dead," he said, with an enormous smile on his face. "But this is great stuff—it's the most important cosmological discovery in fifteen, twenty years. It shows that the universe did evolve through gravitation. Their data fit the Harrison-Zel'dovich-Peebles spectrum, and the amplitude of fluctuations is about right. I'm not depressed. That's what we do—construct theories and test them. Sure, I would have been happier if texture had been proven right. But it's really exciting to see it proven wrong, too. It's great to have real data that can constrain our theories."

I wanted to catch Lyman Page as well, to ask him how close he was to a detection and I caught a glimpse of him at one point, standing around talking with a significant fraction of the world's experts on microwave background detectors—Smoot, Phil Lubin from Santa Barbara, Steve Meyer from MIT, and several others, but he disappeared soon afterward. Having rushed to get down to Washington that morning, he was now rushing to get back. His old mentor from MIT, Ray Weiss, the man who took on "problem cases," was at the APS meeting for the COBE announcement, but would be giving a talk at Princeton later that afternoon on an advanced gravity-wave detector he was involved with, and Page wanted to ride back with him on the train that left at noon. I caught him the next day.

Page had ditched the jacket and tie he'd dug up for the day before and was wearing a lumpy, rust-colored sweater. He spoke in kind of stream of consciousness. Sentences changed direction abruptly. He said, "What we're finding—we're playing it conservative. We're doing the final checks this week and next. If you're going to say you have a detection, you really have to be able to say you're ninety percent confident. We're not there yet. The most probable interpretation is that there is a signal. But our result is also consistent with the signal being zero. With COBE, their result is . . ." He paused, realizing that he was about to twist himself into a grammatical knot, tried to think of a way out of it, and then gave up, ". . . more not consistent with zero than ours."

Page looked out the window up toward the sky and raised his arms as though trying to encompass it. "Inflation predicts a whole panoply of fluctuations, with the same amplitude at all scales. The neat thing is that at large angular scales, you're looking at the primeval spectrum. The fluctuations haven't been processed by gravity or anything else. At smaller angular scales, on the other hand, they're processed. So when you look at small angles, you can see what the processing actually was. It's . . . wonderful . . . it's . . . just . . . the results are going to start coming in fast and furiously now. That inflation predicts this is just beautiful."

He stopped, took a deep breath. "Okay, getting back to this stuff, this is the stuff we've been working on. The COBE result is that the fluctuations are seventeen microkelvins, plus or minus five. We get an upper limit of seventeen microkelvins. Our upper limit is right in the middle of their range. We're looking from three point eight degrees on up, so we'll be able to add information. The prediction from inflation is that the fluctuations should start to have more amplitude at smaller scales because there was more astrophysical processing. If we don't see the power increasing, it'll mean something.

"As far as the theoretical implications go, I think it's still open. That's great for us, since we're still measuring stuff. Now that there's a level though, we won't get this 'You guys will never find any fluctuations' stuff from the funding agencies. This is such a . . . key to understanding what's going on, what's the . . . I mean, it's . . . I imagine that in ten years, we'll have maps of the sky down to one degree, plus people will start correlating the maps with every other map of the sky. The first thing for us to do is cross-correlate with COBE." (A year later, his group would finally finish analyzing the balloon data. Their results would confirm the COBE discovery.)

Tony Tyson had mentioned a similar idea when I ran into him in the hall right after the press conference. He wanted to compare the COBE map with a map of the faint blue galaxies—presumably the oldest galaxies. He figured that it would be interesting to see how structure in the organization of his galaxies reflected structure in the microwave background (of course, he would need an enormous telescope and an array of giant CCDs to do it). I mentioned Tyson's idea to Page. "Yeah, for example. Hmmm, let's see . . ." He stared off into space. "Tyson's looking at . . . boy, I know a good experiment to look for that . . . hmmm, I really have to talk to Dave . . . the Green Bank telescope would be good for that. That would work. That would be a neat thing to do."

He explained that the new 300-foot radio telescope at Green Bank would be the right size for looking at modifications in the microwave background at the same level that Tyson's CCD images of the faint blue galaxies look. "We were just down in Green Bank," Page said, "working at the 140-foot trying to measure anisotropy. I think we need a few more tricks before we can do it, though. It was great. We had to climb right up to the focus of the dish to install our detectors."

I asked him about some of the hyperbole in the COBE announcement—Smoot's statement, for example, that he was seeing God. "Well, I guess Smoot could have been more introspective. I had a phone conversation with someone last night; we just exchanged Smoot quotes. Calling it the 'biggest discovery of all time' the way some people did is a little bit ridiculous. Oh well, it certainly is neat. It's a giant step in our understanding of the universe. For me, one of the great things is that it may change the emphasis at NASA, free up some more money for science and less for the space station. That would be a great thing, since it would get more people doing research. All these people doing experiments are pushing back the limits of technology."

Smoot wasn't the only one who went hyperbolic about the COBE discovery. Both Michael Turner and Joel Primack showed up in sound bites on national television. Turner called it "the Holy Grail of cosmology" (physics and cosmology has been undergoing Holy Grail-itis in recent years: the top quark, the Higgs particle, the theory that ties together all the forces of nature, and high-temperature superconductivity have all been awarded Holy Grail status, mostly by science writers groping for metaphors). Primack floated the idea of Nobel Prizes all around.

For a reality check, I telephoned two of the less doctrinaire astronomers I had met over the past year. The first was George Blumenthal at Santa Cruz. "To tell the truth, I was a little cynical when I first heard the news. I expected that they'd see it, and there it was. In a way, I would have been more excited if they hadn't found anything. But okay, I'll admit it: it's a major thing, a very important milestone. In a sense, COBE is a relief, and in another sense it's a surprise," he said. "It's a relief, obviously, because we really needed to see structure at this level of ten to the minus five if our basic models are right, and we did. It's a surprise because all along, the microwave background has given surprising results, and to see what we expected is so unusual.

"In the sixties, the original discovery was unexpected, though Dicke was looking for it. All the experiments looking at the short end of the spectrum were finding deviations from a black body, right up until the Berkeley-Nagoya rocket experiment. Then COBE showed that it really was a black body. Then there's the isotropy. When the microwave background was first discovered, people should have been surprised when it was isotropic at a level of one part in ten because of the horizon problem. The fact that they didn't see fluctuations should have pointed people in the direction of some mechanism that would have let them be in contact early on, like inflation.

"We should have seen fluctuations in the background of about one part in ten to the three or four, given the fact that galaxies exist today, but we didn't. We also should have expected a dipole anisotropy showing a motion of two hundred kilometers per second, and when it turned out to be six hundred, that was a surprise—we hadn't counted on bulk flows. The former led to many cockamamie schemes, one of which, which wasn't so cockamamie after all, was the idea of dark matter. That brought the fluctuations down to a level of ten to the minus five. And sure enough, there they were. Now we can start ruling out or ruling in models."

My other reality check was John Huchra, who is skeptical of most theories and many experimental measurements as well.

"Look, I know those guys on the COBE team," he said, "and they're real careful dudes. I believe they really measured something. The result is really interesting. But you have to take the interpretation with a grain of salt. Are we really sure these are primordial fluctuations? Are we sure it isn't something closer, along the line of sight, that's pretending to be fluctuations in the microwave background? You're going to get an answer in five to ten years, but not tomorrow."

Finally, I had to touch base with David Wilkinson. For him the COBE result was a fitting monument to the later part of his career, just as the search for the original microwave background and the first planned detection of it defined the early part. We made an appointment, but when I showed up in his office at Jadwin Hall four days after the COBE result was announced, he wasn't there. He appeared a moment later, walking down the hall with a bow saw in his hand. Even with all the hammering and wire cutting I'd seen in Green Bank, I couldn't imagine what kind of physics he was doing with a saw.

"Oh, that's for a course I'm teaching to nonphysics majors," he told me. "One woman was trying to calculate the fractal dimension of a tree [fractal dimension is a quantity that describes the complexity of objects like trees, mountains, clouds, coastlines, and, a few scientists think, the universe, which have structures on many scales]. She started out by taking photographs of the tree. But then she started worrying about whether projecting a three-dimensional tree onto two-dimensional photographic paper was giving her false results. So she decided to cut up a branch and weigh the various pieces, and that's why we're using a saw. This course is really fun to teach—there are only four kids in it, so they get a lot of attention. And they really get excited about the science. They're doing good work."

He leaned back in his chair, cupped his hands behind his head, and smiled. "Well, COBE worked. I didn't think it would. We actually first saw this signal back in September. Ned Wright was the one who got it. We had all these procedures set up for analyzing our data, but Wright is one of those young guys who's great with a computer. He said, 'Just give me the data,' and he did an end run around the rest of the team and saw a signal. The trouble was that we didn't have the systematic errors low enough, so it wasn't a sure thing. We had to worry about effects like the Earth's magnetic field, interference from our own transmitters, interference from Defense Department radars, which are naturally interested in things orbiting overhead. If those things lock on to you, you can have real problems: geosynchronous communications satellites above us, thermal effects from heating by the Sun—that's about half of them.

"So what we did from October to January, and what's still going on to some extent, was to try and eliminate all possible sources of error. Once we convinced ourselves it was nothing else but a cosmological signal, we started writing the papers. At this point, I feel pretty good about the result. In September, I wouldn't have given much of a chance that it was right. I never thought COBE would get down to this level of sensitivity. It is the design level, but that's based on the irrational idea that we wouldn't run into any systematic effects. In other experiments, there has always been some unexpected problem. I'm surprised and pleased—it's been twenty-five years. Even while we were slogging along on top of Guyot Hall, doing the first experiment intentionally designed to measure the microwave background, we were thinking of ways to measure anisotropy. We missed getting the dipole by a factor of two, and then Smoot and his group put a radiometer on the back of a U-2 and saw it clearly. Things have been inching down since then, and we've finally seen the quadrupole anisotropy at a level of six times ten to the minus sixth. At ten degrees, it's ten to the minus fifth. The action is at one degree, where the things like the Great Wall and the Great Attractor are going to show up.

"The result agrees with inflation—doesn't prove it—and rules out texture. Peebles said yesterday that his PIB model is still alive. CDM has survived, although there's a problem with the bias parameter, HDM—I really think there's got to be something to it.

It's so natural. You know there are neutrinos out there, just give them a little mass and relax.

"Anyway, the neat thing is that COBE's up, and it's still working, and more data processing will pull the errors down. We still have more noise than signal, and so the detection of hot spots is statistical—you can see spots on the false-color images, but they're mostly not real. You can prove it by blinking between two maps at different wavelengths. I did that at the talk we gave up here, and the spots appear to jump around. Real hot spots wouldn't do that."

At that moment, a boyish face poked around the corner. It was Richard Gott, come to call on Wilkinson. "Oh!" he said, seeing me. "I guess I know why you're here. If you want my opinion, this is a watershed, a major milestone, a triumph for the theory. It really vindicates the whole CDM model. I've been saying this all along, of course." Wilkinson couldn't resist. "Now Rich, you have to admit that a bias factor of two still causes problems."

What he meant was that despite what Smoot and Wright had said in Washington, and what Gott was saying now, COBE had not necessarily vindicated the standard CDM model of structure formation. One of CDM's necessary tricks was to assume galaxies were grouped together more tightly than dark matter in general; their location in the universe was biased. The trouble with the COBE results was that the most likely value for the bias factor, according to the theorists, was somewhere around two, meaning the galaxies were about twice as tightly clustered as dark matter.

But COBE's hot spots, presumably perfect mirrors of the clumpiness of matter at the very beginning, pointed to a bias factor of one; that is, no bias at all. COBE's structures were on the very largest scales, but an extrapolation to smaller scales seemed to imply that ordinary matter and dark matter were clumped identically. That in itself wasn't a problem, but it turned around the long-standing argument against CDM that there was too much structure on large scales. Large scales would now be okay, with Great Walls and Great Attractors falling nicely within the theory, but now there was too *little* power on smaller scales. There should be more structure in clusters of galaxies than there actually was.

"Oh, sure, of course," said Gott. "I'm not saying there isn't some tough work ahead. No doubt about it. But of course this result is on scales where you don't have to worry about all these problems. Sure, hydrodynamics is important, and galaxy formation and all, but this is a number we can peg our hats to and work down from there."

"There will be a great parameter adjustment," agreed Wilkinson.

There then ensued a lopsided discussion between Gott, the front-porch storyteller, and the unflappable Wilkinson. Gott strode back and forth, waving his arms, opening his eyes wide to emphasize a point, his voice rising and falling dramatically. Wilkinson listened politely but managed to break into the sermon with his quiet, reasonable voice.

"People always ask, does inflation have any real predictions," said Gott, "and this is the one, the flat power spectrum. That's what inflation does for you that no other theory does."

"Well, there's also the fact of thermal equilibrium all over the sky," said Wilkinson. He was talking about the horizon problem, inflation's other great triumph.

"Right, right, but remember that that was already known before—this was a real prediction."

"Well, of course, the flat power spectrum was around before, too."

"Sure," said Gott, "the Harrison-Zel'dovich spectrum."

"You know, Rich, you guys should go back and read those papers. You'd see that Jim Peebles also figured out the flat spectrum back in 1970."

"Okay, well, fine, that's fine. If Peebles said it, he should get the credit. No question about it."

Again, they had sounded a warning note on the COBE results. The fact that the pattern of hotspots was similar to the Harrison-Zel'dovich-Peebles power spectrum didn't for a moment prove inflation had ever happened. Inflation was perhaps the best theory that could explain the power spectrum, but the three cosmologists had argued long before inflation came along that the flat power spectrum was the only one that was reasonable. Besides, the power spectrum COBE had seen was not airtight. It was reasonably close to the Harrison-Zel'dovich-Peebles spectrum, but not right on it.

"Now," said Gott, "you have to realize that while this is a real triumph of the theory, there's still a lot of mud wrestling we have to do. As someone once said: 'If physics is a chess game, astrophysics is like mud wrestling.' We have to make this result consistent with a lot of other results, all the other data."

"Now you've got to be careful about making your claims too strong, Rich. This doesn't *prove* inflation. But if you got ten cosmologists together in a room, I'll admit that they probably would have predicted a power spectrum just like what COBE saw."

"The thing I found very impressive at your talk the other day," said Gott, "was when you blinked those pictures back and forth. It really showed how much of that signal is real, the way some of those hot spots didn't seem to move from one picture to the next."

Wilkinson smiled. "My point was that they *did* move. I was trying to show that they were noise."

Gott looked blank. "Oh. I thought . . . oh. I thought the point was that they didn't move. I see. Oh, well, I'm just an optimist."

Over the next few weeks, Gott's statement began to look prophetic. As the euphoria of the COBE announcement faded and people started looking more closely at the data, it became clear that nothing had really changed very much. Important as it was, the COBE result didn't actually prove or disprove anything; it was one more piece of data that cosmologists could fit into their theories, and what they saw depended largely on what they already believed. People like Gott considered CDM all but proven by COBE, but he had thought it was proven before; people like Neil

Turok concluded it was all but ruled out (contrary to what David Spergel had said in Washington, Turok was standing steadfastly by the theory of cosmic texture).

In the month or two after the APS meeting, I had two chances to watch astrophysicists try and figure out what the COBE discovery really meant. The first was at an informal lunch at Lee's Castle, a Chinese restaurant that's an exception to the rule about there being no good ethnic food in Princeton. The participants were Lyman Page, Ed Turner, David Spergel, Neta Bahcall, and a young theorist from the Institute for Advanced Study named David Weinberg, who has worked on some of the most advanced computer simulations ever made of the evolving universe. Playing just a little dumb, I asked them all how important the discovery was.

David Spergel  PHOTO: EILEEN HOHMUTH-LEMONICK

DAVID SPERGEL: I think it's rather important. My first reaction was that it's telling us that we're in a sense going in the right direction by imagining that structure formed gravitationally. The amplitude of the potential fluctuations, if you assume these *are* potential fluctuations that COBE was seeing, is consistent with the amplitude of galaxies, so that . . . well, it means that Jim Peebles's book on large-scale structure is worth owning. There was always a distinct possibility that structure formed some completely different way, and I think the fact that we're seeing structure on such large scales suggests that we've been thinking in the broadest sense about the right kind of picture for its formation.

LYMAN PAGE: So you're saying structure probably came from gravity rather than, say, explosions?

SPERGEL: Yes, or radiation forces or some of the other things people have come up with.

ED TURNER: It shows us that gravity was important, not necessarily that nothing else was.

SPERGEL: Right. It says gravity matters.

DAVID WEINBERG: It tells you that the fluctuations were there, and were there at the level that, if you take the best existing theories, they ought to be to produce more or less the structure that we see within a factor of two. What we're seeing is what a theorist's best guess for what we should be seeing would be, except that I was . . . I just didn't expect that the theories would be right.

TURNER: Yes, that's the thing about it. You don't know whether to be impressed or depressed by the fact that it seems to correspond to the most conventional, straightforward theoretical expectations, which in some sense really do involve big extrapolations, both in space and in time. It's always amazing to see something in the universe now, and then be able to check it with some observation from the very early universe.

NETA BAHCALL: So if we look just at this one, single observation plus the upper limits from other observations, what theory is ruled out?

David Weinberg   PHOTO: MICHAEL D. LEMONICK

SPERGEL: Standard CDM is ruled out.

BAHCALL: Standard CDM, right.

WEINBERG: I don't actually think that's true.

PAGE: What do we mean by standard CDM?"

BAHCALL: It means some sort of bias—the matter is not distributed like the galaxies—omega of one, and the scale invariant spectrum. It's ruled out because if you take the COBE result at this very large scale, and then you extrapolate it with that spectrum all the way to the small scale of galaxies and clusters, then the galaxies should be moving around faster than they are by a factor of more than two. These are small-scale flows we're talking about, not bulk motions. In that sense, that's what Ed means about not knowing whether to take it as a success or as a failure. On the other hand, it is quite impressive that you get a spectrum very close to what CDM wanted. Of course, Harrison and Zel'dovich and Peebles suggested that spectrum independent of any theory—it came before inflation or CDM—but CDM also gives that spectrum. Now if you change that spectrum to just a slightly different one, it would be consistent with COBE and maybe consistent with other observations. So the CDM people would take that as a success and say, well, we're very close.

The thing is that you look at the COBE results: the problems you find are exactly the same and in the same direction as the problems you already had with CDM from the large-scale structure observations of things like the clustering of galaxies and of clusters.

TURNER: In other words, you can ask how COBE changed the question of whether standard CDM is right, and the answer is that it didn't change it at all: we already knew the probability was zero that it was right, and it still is.

BAHCALL: But some of the theorists were saying, "Well, those large-scale structure observations are wrong, and if you change just a bit in the right direction, everything will be okay." They will now have a harder time doing that.

WEINBERG: You have to remember, though, that COBE's measurements have pretty wide error bars. If you take the best fit to the COBE numbers, I'll agree there's a problem.

BAHCALL: I think if you take the best fit, the only thing you can change in CDM is the spectrum of fluctuations.

WEINBERG: No, you could have velocity bias. [That is, you could argue that just as galaxies may be biased tracers of mass, they could also be biased tracers of the motion of matter in the universe—the dark matter could somehow be swirling around faster than the galaxies are.]

BAHCALL: Velocity bias? I wouldn't buy it for ten dollars. I believe the observations.

TURNER: One of the arguments for evolution is that you find fossils in the ground, and the creationists respond, "Well, God put those there to trap sinners." In a sense, velocity bias is in this same spirit. It's kind of like a conspiracy theory.

BAHCALL: The critical thing is that spectrum, right? All along it was, with cold dark matter you need to have a scale-invariant spectrum of perturbations. Now I suddenly hear, well, it doesn't *have* to be scale invariant, because cold dark matter doesn't *really*—where does it come from? Does it come from the cold dark matter, or does it come from inflation?

SPERGEL: Look, there are really two things going on here, right? First you have these initial fluctuations, and then you have what we call a transfer function, a rule that tells us how the fluctuations are modified over time. And that transfer function depends on whether it's cold dark matter or hot dark matter or baryons. Then you need some initial source of fluctuations. The initial source of fluctuations . . . well, in the simplest inflationary models, you're going to have one set of fluctuations. If you modify inflation, then you get a different set. In either case, you then have to apply the transfer function.

BAHCALL: So it's the combination of inflation plus the cold dark matter transfer function that gives this standard model. If you change the transfer function, it's not standard CDM anymore.

WEINBERG: Right. For example, maybe there's cold dark matter, but omega is low. Or the Hubble constant is very low.

SPERGEL: If the Hubble constant is thirty-five, it would just solve everything. Cold dark matter would work brilliantly.

WEINBERG: Or that there are more neutrinos than we know about.

BAHCALL: I love these discussions between theorists and observers sometimes. Theorists will go complain to the observers: "You guys are giving us these factors of two," and then the observers get after the theorists about all these models and all these infinite parameters. Of course, if you tweak enough things enough times, you can get it right.

PAGE: The large angular scale observations are hard to get around, though. These sections of the universe were causally disconnected, and that's just got to tell you something primordial.

TURNER: But we have this general trap of trying to live between this clean early regime, which I think is pointing one way, and a lot of late-time, low-redshift, messy observations, which you can kind of doubt. I mean, if you take the observations seriously, you have to live with a fairly high value of Hubble constant. There's a fairly good case for that. And there are the large-scale flows, and on and on. Quasars that form at high redshift. So on one end it's very clean, and on the other end it's very messy, on both the theoretical side and the observational side. Maybe the large-scale flows aren't really real, or there could be all sorts of problems with H-nought. I have, and I think most people have, kind of an uneasy feeling that what works with the microwave background fluctuations on large scales looks like it's not going to work at low redshift at small scales.

WEINBERG: Right. You don't know which observations are wrong and you don't know which theoretical computations—not the assumptions, but the actual computations—are wrong.

BAHCALL: But if you look at it in an optimistic rather than a pessimistic way, you can say, boy, it all fits together within a factor of two or three, and that's quite amazing.

SPERGEL: I disagree with Ed. I kind of have the feeling that it almost fits, and that we're maybe one step away from figuring it all out.

WEINBERG: But it's hard to know whether it's a small step or a giant step. I mean, Newtonian gravity just about almost always works. It almost gets the precession of Mercury right. But the fact that it falls short is tremendously important.

And you also don't know which part of the theory you have to change to make things fit, or whether it's a minor change or a big change. You want the observations for guidance in there, but because a lot of these are very tricky observations, you also don't know which are ones we have to fit and which are ones that we shouldn't be so worried about. And in terms of the theory, the problem is . . . well the quasar thing is a good example. We know very definitely that there are quasars at a redshift of four, so a model has to explain them. But we understand so little about how quasars form. You can say, well, this theory is going to have a harder time than this other one at making high-redshift quasars, that it's really going to have some trouble. But to make that case you have to be able to say that you know what will happen with some very messy astrophysics. And we're really not in that situation.

TURNER: And David's right that it's also not so easy to calculate from the theories. For example, some of the recent CDM simulations have shown that they make things earlier than people expected.

SPERGEL: One of the things about this that's surprising to me in a sociological way is that all the calculations people are now thinking about doing, now that COBE has seen something, could have been done five years ago.

But they weren't done. And all of these questions were interesting before the COBE results, but only now is everyone really thinking about them and excited about them.

WEINBERG: It also changes the significance of existing papers. There's this primary reference, by Jon Holtzman. And, you know, I remember when I first saw that—he just took a whole slew of different models and calculated them—I thought, that's not that interesting. But now that you have the COBE numbers, I go back and say wow, I wish he'd done *this* model, where omega equals point four and the cosmological constant equals point six, or whatever.

TURNER: I think never mind how structure formed. It's interesting and important, but I mean, structure formed somehow, we know that, and it had something to do with gravity. It's still a very interesting problem, but I really think it's a more fundamental problem to try and understand what the curvature of space is, whether the universe is open or closed or flat.

WEINBERG: I can well imagine that eventually we'll look back and realize that the most important thing that this observation established was that the universe was flat, even if we didn't realize it at the time.

PAGE: From the observational point of view, the important thing is just that there is anisotropy. There are now a whole lot of systematics you don't have to consider as strongly now. Now you know there's something there, so you have a target.

BAHCALL: If we can get that answer, if the COBE result could give a really definitive answer that the universe is flat, that would be very important.

TURNER: I think it's a little embarrassing that we don't know the answer to that question theoretically. I mean, talk about interesting calculations.

WEINBERG: Except that as Rich Gott says, on the one hand it's more important to know the result, on the other hand its even less fun to do the calculation, because it's sort of messy and difficult.

SPERGEL: Actually, I think it's much more interesting now. Most of the theoretical attention has been focused not on the microwave background but on large-scale structure. Dick Bond was giving the one microwave background talk where he put up all those curves with upper limits from different experiments. And limits are important, but . . .

WEINBERG: But they're boring.

SPERGEL: Let's face it. The microwave background was important but boring. Now it's exciting.

BAHCALL: But I would have thought that once we got a number we'd know where we stand, and it's not quite true because the theorists start tweaking the parameters, and then you can fit anything.

WEINBERG: The other thing is that the "point" that COBE found still has an error bar of a factor of two. If this were the only result we were ever going to get, we'd be disappointed, because it's tantalizing, but not enough information to get you solid answers. The great thing is that it means this field has started observationally, and in a few years we really should have much better numbers.

TURNER: On the observational side, I think I don't know any group of people I'd be more inclined to trust than the COBE team, but Dave Wilkinson gave this extremely sobering talk in Irvine, where he explained how complicated it was to analyze the data. The nice thing is that they have a map, with hot spots, and other people can look for them. But . . .

PAGE: But they're hard to find.

TURNER: Right. I mean, is there anybody at all worried that such a complicated data analysis was done by people who knew what the answer ought to be?

PAGE: I'd worry about the method of analysis. I mean, the quadrupole is going to stay there. But that other number might move. What we're sort of getting right now is fourteen, fifteen.

TURNER: Fourteen or fifteen what?

PAGE: Oh, sorry. Microkelvins. I'm so used to talking about this with other microwave people.

BAHCALL: You're getting fourteen compared to their, what, seventeen?

PAGE: Right. But it may move up or down. It's so tricky. My guess is that their number is going to come down slightly. But I think they're going to stay within their quoted errors.

WEINBERG: Do you think there's any possibility that the results will completely go away?

PAGE: No.

WEINBERG (LAUGHING): Good. On what basis? On the basis that they're very careful, or that other experiments are seeing it as well, or . . .

PAGE: No, I think just that they're careful. I mean they saw the quadrupole very early on, within a month after COBE went up, and nobody believed it, so they didn't publish it.

WEINBERG: But you don't think it will turn out to be the galaxy.

PAGE: It's not the classic galaxy. It could be some funny dust cloud, some halo around the galaxy, that has some funny opacity. You have to invent things like that. That's my impression, just talking to those guys. And they're really topnotch, the guys who are pounding on this stuff. Dave Wilkinson is pretty hard-nosed about eliminating systematic errors. And Ray Weiss, he's a pretty cautious guy. And George Smoot, too.

TURNER: I think we should point out that there's some history in this field of discoveries that did go away.

WEINBERG: Right. I found it sobering when I was looking into this open universe question, and I went back to a couple of papers that were written in '81 and '82, and they were talking about the detection of the quadrupole at the level of ten to the minus four by Wilkinson et al. I sort of read that, and went, whoa!

TURNER: True. This is the second time it was detected.

PAGE: Right, and that time it turned out to be the galaxy. But the advantage of COBE is that it's multifrequency and full sky. You cut thirty degrees away from the plane and it's still there. The only way this could be our galaxy would be to make up something crazy, some halo with dust in it.

WEINBERG: If this scale-invariant spectrum holds up in the observations Lyman and these other guys are making at small angles, it will really be amazing to realize that Harrison and Peebles and Zel'dovich could make these very simple arguments in 1970 that, well, this is what it ought to look like. It was already impressive when people came along with real, physical theories for producing the fluctuations, like inflation, and the theories predicted what they had come up with. And now it looks like the universe agrees.

TURNER: That's the curious thing about the universe—for the most part, people often say that physics is beautiful because it's simple. But what's really interesting is that the laws are simple but that by and large, the realizations of those laws are complicated. We have an incredibly complicated world coming out of some very simple laws. It seems like an odd and special property of the universe that it does seem to like the simple possibilities. Turning that around, does this really give us evidence for inflation?

WEINBERG: Yes, we may find ourselves in this really funny state where we can confirm inflation as well as we can hope to for the foreseeable future and yet not know whether we've given it a strong confirmation or not.

TURNER: To say you've proven inflation is to say—basically you're really saying that you predict no one will think of a better idea to explain the observations.

SPERGEL: The disturbing thing about inflation is that you can't disprove it. Even with low omega, there are modifications you can make.

WEINBERG: Well, I'm not sure there's any believable model of inflation that has low omega and doesn't also give you excessive microwave background fluctuations. That may be a falsifiable aspect of the theory.

SPERGEL: I don't know. I would bet that if you came up with definitive evidence that omega was low, someone would come up with a model that would make it work. It's a slippery subject.

About a month after the lunch, I got a call from David Spergel. He and some other astrophysicists at Princeton, realizing that the COBE announcement had raised more questions than it answered, decided to pull together a small, informal conference of cosmologists from around the country to try and get a handle on the theoretical and observational situation. What exactly did the result mean, and how did it fit in with other data?

Considering the short notice, it was a testament to the Princeton department's prestige that so many top cosmologists showed up for the meeting. Smoot, Wright, and the entire COBE team were there. So were Nick Kaiser (the father of biasing) and Dick Bond (the namer of hot and cold dark matter) from Toronto. Phil Lubin, who was leading the race to detect small-scale anisotropy, came from Santa Barbara. Bernard Sadoulet, on his way to a meeting in France, stopped by. Marc Davis, John Huchra's partner in the early Harvard-Smithsonian redshift surveys, and George Efstathiou from Oxford were there. Together, they made up one half of the original Gang of Four. Alan Guth, the father of inflation, came down from MIT, and Paul Steinhardt, one of its uncles, came over from the University of Pennsylvania. Robert Wilson, who found the microwave background in the first place, came from Bell Labs. The local talent came out in force as well: Peebles, Ostriker, Wilkinson, Spergel, Ed Turner, Neta Bahcall, Page, Spergel, Turok, Gott, and Jim Gunn all showed up.

The astronomers began assembling at about eight-thirty for bagels, coffee, and registration. Then everyone filed into the Peyton Hall auditorium, where Spergel opened the meeting. "For those of you who want to send or receive E-mail, we've got a terminal set up in room one twenty. The password is 'Zel'dovich,' with the proper spelling. If you can't spell it, you can't log on."

Then the sessions began, on topics like "Microwave Background Experiments," "Implications of COBE for Cosmology," "Observations of Large-Scale Structure," "Does Anything Work?" and "Where Do We Go from Here?" Originally, Spergel had figured three or four speakers per session, and lots of time

Astronomers at the Princeton COBE workshop, June 1992

PHOTO: EILEEN HOHMUTH-LEMONICK

for questions; it was to be more of a workshop than a conference, letting the astronomers challenge each other and try to
work toward some sort of consensus. In practice, though, the
three or four grew to eight or ten per session. No one wanted his
or her two cents left unmentioned. The result was somewhat
breathless. Over two days of the workshop, just two speakers
finished their talks on time. The rest had to be intimidated off
the floor by the session chairperson, who would stand up and
hover menacingly until the speaker, rushing through the last few
slides, gave up. There was hardly time for any questions.

The questions that got asked were unusually pointed, though.
In the two months since the Washington announcement, not much
had changed. None of the speakers had given up on his or her
dearly loved cosmological model in the wake of the COBE detection, and no one in the audience was willing to let someone else
get away with pushing a competing model too hard. Those with
no particular model to push concentrated on their favorite themes.
David Wilkinson, who chaired the first session, stood up and
announced that it was time for his annual wart talk—warts being
a Wilkinsonism for the problems that can fool scientists into
thinking they're seeing a signal (World War II flyers would have
called them gremlins). "A wart," he explained, "is small, but it's
ugly."

Then George Smoot and the COBE team got up one by one and
went over their results again. They largely rehashed what they had
said in Washington, although Smoot did present a new idea to
explain why the hot spots COBE found weren't precisely in the
pattern standard CDM would have predicted: there's a good
chance, he said, that gravity waves (the same kind of waves Tony
Tyson had looked for in the early 1970s) were generated during
inflation, and these, according to both Smoot and Dick Bond,
could have messed up the spectrum of fluctuations. This created a
small ruckus, with Jim Peebles getting into an arcane discussion
with Bond, which no one else could follow. (They continued the
argument later that evening at Dave Wilkinson's house, their exag-

gerated gestures and careful enunciation a testament to their host's generosity with his cognac. "I will wager a bottle of the finest New Jersey wine," said Peebles, "against one of your best Ontario wines, that you are incorrect.")

And then the cosmologists began to parade by, each explaining why his or her interpretation of COBE should be taken more seriously than anyone else's. The most entertaining by far, everyone agreed, was Richard Gott, whose rapid-fire talk was filled with dramatic asides and viewgraphs shuffled on and off the projector just a little faster than anyone could register them. Gott referred back to the time of Copernicus, whose Sun-centered cosmology, published in the sixteenth century, displaced the Earth-centered Ptolemaic universe that had been the standard model for nearly two thousand years. Although Copernicus was essentially right, the details of his cosmology were quite wrong, and that, said Gott, was worth remembering. "The amazing thing," he said in defense of CDM, "is that you shoot an arrow fifteen billion years back in time, and whack! You hit the target, but you miss the bull's-eye by a little bit. Is this a failure of the theory? I say it's a big success. Cold dark matter is the Cheshire Cat of cosmology [he projected an illustration from *Alice*]. It's invisible, but here's a picture of its smile." He flashed a photograph of a cluster of galaxies with an enormous Tysonesque arc, a lensed background galaxy, which looked like an enormous grin. "The big question is not is the model correct in all its detail, but is the universe made mostly of nonbaryonic matter, and"—he looked over his shoulder at the arc on the screen—"is it laughing at us?"

The rebuttal was delivered by Neil Turok, a little bit defensively. "A number of people speaking here today have oversimplified the situation. We keep talking about cold dark matter, but that isn't the only theory we have. Inflation is a beautiful idea, but no one knows what inflation came from. I've been working on this other theory, cosmic texture, not because it's true (and I would say that neither it nor inflation is likely to be true in the end), but because it is equally plausible and equally beautiful. Has COBE confirmed standard cold dark matter and inflation? I'm going to be controversial here: no. First, the anisotropy isn't quite right, unless the Hubble constant is fifty and the bias parameter is one. Second, the power spectrum agrees with inflation but it also agrees with every other theory in town. I'll admit that texture, unlike inflation, doesn't address the flatness and the horizon problem. You have to take flatness and the isotropy of the background radiation as part of the initial conditions of the universe. But inflation isn't any better at explaining initial conditions. It's just good at erasing any evidence of them."

Finally, as the day ended, Spergel claimed his right as the conference organizer to ask a final question of the hundred or so astrophysicists, physicists, and graduate students in the room. "Can someone tell me an observation that could falsify inflation?" Hands shot up. "Space curvature," said Jim Peebles. "Omega significantly less than one." "Bulk flows on very large scales," said Marc Davis. "A totally incompatible spectrum of fluctuations," said Paul Steinhardt. "Evidence that the universe as a whole is rotating," said Smoot. "You know," said Steinhardt, "bad theories don't usually die. They fade away." "Right," agreed Guth. "Theories are discarded not because they're falsified, but because better theories come along."

With that observation, the day's meetings ended and the entire group retired to—where else?—Lee's Castle, for a Chinese banquet and dozens of miniworkshops over shark's-fin soup and sautéed lobster. After that, it was over to Ed Turner's house for coffee and relaxed conversation, or Dave Wilkinson's, for cognac and arm waving. The second day was much like the first, with no existing theoretical model abandoned and several new ones proposed (Ostriker suggested something called tilted cold dark matter, Nick Kaiser brought up velocity bias, and George Efstathiou suggested adjusting CDM by assigning the universe different bias factors on different scales). Every statement was subject to loud challenge.

The final speaker was Ed Turner, as always skeptical of the

conventional wisdom and as always able to express himself so good-naturedly that it was impossible to take offense. He put up his own listing of what the new COBE observations proved: the Big Bang happened. The universe inflated. Standard CDM is right. Standard CDM is wrong. There is no bias factor. The bias factor is two. The universe was reionized. God has a face. Cosmologists tend to overinterpret data.

"Of all these," Turner said, "I am only certain about the last one. I think the moral I can draw from what I've heard over the past few days is that being close to the truth but wrong might mean we're close, but it might mean we're wrong. Those who look to history for analogies will often be misled. And I think we should at least consider trying other approaches to the problem of structure formation than merely patching the standard model with epicycles." [Epicycles were the cumbersome orbital adjustments added to the Sun-centered Ptolemaic cosmology to make the theory fit the actual motions of planets.]

The very last event of the conference was a cookout on the wide, tree-shaded lawn behind Dave Wilkinson's house, down by Princeton's picture-postcard artificial lake (it was created by damming a stream, so that the Princeton crew team would have somewhere to row). The air was warm, the rhododendrons were in full bloom, and the sky glowed for what seemed like hours after the Sun had set. It was such a perfect evening that even the Californians were impressed. A few of the astronomers had brought their children—Lyman Page's two-year-old son, William, whose day care arrangement had gotten me an introduction to his father, toddled around the lawn; and Jim Gunn and his wife, Jill Knapp, another Princeton astronomer, had brought the brother and sister they'd adopted from Peru a year earlier. The bigger kids wrestled with Wilkinson's Labrador retriever, while Wilkinson and Page blew giant soap bubbles for William. The grown-ups wandered down to the lake or sat on the patio, listening to the crickets singing.

It would have been impossible in such a setting for any of the day's cantankerousness to survive. The cosmologists were feeling relaxed and reflective, and as they ate their grilled chicken and asparagus and strawberry shortcake, I drifted from group to group and asked several of them what they thought of COBE and of this first high-level conference to study its results. "What's exciting here," said Turok, "is that we really *don't* understand what COBE is telling us. The measurement has given no sense of confidence about what's going on out there. I think it's likely, when things have settled down, that this result will be remembered for having ruled inflation out."

"What we really need," said Peebles, "is another season at the South Pole, and more microwave observations on smaller scales. Virtually all of the models mentioned today were variations on conventional CDM. I really wonder whether someone will come up with an elegant alternative theory. I may be getting crotchety, but I think people are doing too many big computer simulations, and not enough thinking."

"There's a big danger in taking the COBE measurement as gospel," said Spergel, "which is that you know the answer you ought to get. If you get a result that looks crazy, you stop and try again. But if you get an answer that looks plausible . . . well, you still check, but not quite as hard."

"It's really kind of crazy," said Neta Bahcall. "Here is a measurement people expected, and instead of ruling out models, we've seen more than ever. I still think that the simplest idea may be the right one. Omega is low, and the Hubble constant is seventy-five or eighty—just what the observations are telling us. The sociology of all this is fascinating. Of course, everyone is emphasizing the particular model they're working on. And you fight it out until new observations disprove a model or until a new theory comes along. These meetings and these fights are useful. They get you thinking. They make you see some of the problems that exist."

Finally, it was too dark to tell a CDM enthusiast from a low-omega booster, and I started talking with a Dutch astronomer, an expert on pulsars who had audited the workshop to see what the cosmologists were talking about these days. He had found all the controversy very amusing, but he was more fascinated at the moment by a phenomenon unknown in the Netherlands and also in the mountains of Arizona. The stars would be at their best that night, which meant, this being New Jersey, that they would be few and dim, but the sky was blazing with fireflies. We went out on the lawn, into a local universe filled with tiny Cepheid variables, and I began showing him how to sneak up and catch them.

EPILOGUE (1995)

It is usually impossible to tell when a scientific revolution is complete until well after the fact. A model that purports to explain some aspect of the natural world—the Ptolemaic solar system, Newton's laws of gravity, uniformitarianism and catastrophism in geology, particle physics before the 1950s—has more and more trouble explaining new observations. The model becomes more complicated and cumbersome, as Ptolemy's did, or it simply ascribes the anomalies to experimental error, the way physicists did when Newton's conception of gravity could not quite explain the precession of Mercury's orbit around the Sun. Only years later, when the new model has explained the existing data and, more important, has successfully predicted the outcome of new observations, does it begin to replace the old. Einstein's general theory of relativity was fascinating and mathematically compelling, but until Sir Arthur Eddington measured the positions of stars during an eclipse and found that Einstein's predictions of the gravitational-lensing effect were more accurate than Newton's, no one was ready to buy Einstein's model. The relativity revolution actually began with the publication of Einstein's theory in 1915, but this only became evident in the years that followed.

The Big Bang model of the universe is perhaps the best example of this phenomenon: it first emerged in the 1920s, with the discovery of the expanding universe, but it remained just one of several competing models for four decades. It wasn't until Arno Penzias and Robert Wilson found the cosmic background radiation predicted first by Gamow, Alpher, and Herman (and later by Dicke), that it was adopted almost universally.

It is also true, in physics at least, that simpler models tend to be more correct than complicated ones. In particle physics, for example, a model based on six quarks and a handful of leptons is more aesthetic than older models that tried to account for hundreds of "elementary" particles. Experimentalists also say that the simpler theory is closer to the truth, but no law says that the correct theory must be the simplest one.

When I last visited with cosmologists in the spring of 1992, just after the COBE satellite had detected subtle variations in the light at the edge of the universe, the field was certainly in the process of a revolution; it was quite clear that no existing model of cosmic structure could explain observations of galaxy clustering, galaxy motions, or dark matter. The cold dark matter model, elegant and apparently on its way to general acceptance in the mid-1980s, was clearly wrong in some ways. The COBE result didn't help much. It didn't quite vindicate any particular theory, yet it didn't quite rule any of them out either—at least according to each theory's most stubborn proponents (for these people, of course, COBE did all but rule out competing theories).

Two and a half years after the COBE result made front-page headlines and dominated the evening news, I wanted to know what the state of the revolution was. Had any of the dust from COBE settled? Was it finally living up to its initial billing of being "like seeing God," as George Smoot had proclaimed, or of being "the most important discovery of the century, if not all time," as Stephen Hawking had said? Had it, or any other important observations since, convinced astronomers to stream, like galaxies moving in the direction of the Great Attractor, toward any one model of the universe?

I went back to some of the people who had led me through the unfamiliar terrain of cosmological theory and practice and asked them what was new. (I did a lot of it by E-mail this time. Cosmology may have not settled on the direction of its revolution yet, but communications clearly has. Like hundreds of thousands of other Americans have over the past few years, I have gotten an E-mail address, which makes contact with astrophysicists an order of magnitude quicker and easier. I could presumably even get John Huchra to forward his bowling team's latest scores, but I have not yielded to that temptation.)

The short answer is that not very much has changed. The people who were passionate about cold dark matter—with its attendant assumptions that the universe underwent inflation, has the right matter density for the cosmic expansion to slow nearly to a stop but never to fall back on itself (i.e., that omega is one), and is permeated with exotic particles—are still passionate. They have been willing to yield on the idea of biased galaxy formation, which COBE challenged in its simplest form. Maybe galaxies still tend to form only where the concentration of cold dark matter is greatest, but that tendency varies depending on scale.

Then there are the people who, in COBE's aftermath, still bought the idea of inflation and the fact that omega is one, but decided that the dark matter must be a mixture of hot and cold. The hot is neutrinos with just a little mass, which would neatly explain galaxy clustering on very large scales. The cold is still old-fashioned CDM, which handles the small-scale stuff. Joel Primack, at the University of California, Santa Cruz, is quite happy with this notion. "It's true," he said, "that the original cold-plus-hot model didn't quite fit in with the COBE data, but that was based on certain assumptions about the neutrino's mass [on our walk through the Santa Cruz redwoods he had been talking about a ten-electron-volt neutrino]. If you drop the neutrino mass to about two electron-volts, though, it does just fine."

The people who didn't like the complexity of the mixed–dark matter model before, however, didn't like it any better now. Princeton's Jerry Ostriker, who had compared it originally with a badly overseasoned pot of soup, was one of these. Last year, he and Renyue Chen created a computer model of a cold-and-hot-dark-matter universe with omega equal to one, and proclaimed that it simply did not work. You could not create the observed structure no matter how you tried. (Primack's comment: "Jerry is wrong. I explained it all to him before he published his paper, and I don't understand why he's still saying these things.")

Along the same lines, those who believed in inflation still did; those who had never been convinced still weren't. One of the most passionate believers in inflation, it turned out, is Stephen Hawking. I had never spoken to him before the COBE result came out for the simple reason that, hard though it must be for the general public to believe, Hawking is not the world's greatest authority on every topic in physics and astrophysics. If you want to talk about black holes or singularities or the more abstruse implications of relativity, Hawking is a key player. On cosmology, he is only tangentially involved. About a year after COBE I was offered a chance to interview Hawking on the occasion of the publication of his second popular book, *Black Holes and Baby Universes,* and I jumped at the chance to meet the world's best-known scientist. Hawking was on the West Coast for several scientific meetings and public lectures, and he offered me an hour or two in Seattle between commitments.

I was fully prepared not to be overly awed by a man who, for all his important scientific work in the face of an increasing (and now nearly total) loss of body control due to amyotrophic lateral sclerosis, is still merely one of the world's top physicists and not a superhuman intelligence. I failed. Perhaps it's just all the publicity he has received, but I really did feel that I was in the company of a presence. He didn't quite glow, but almost. Hawking sat in his wheelchair, only his eyes moving to look from me to his touch-activated, wheelchair-mounted computer. He can speak only through a voice-synthesizer, his voice destroyed by a tracheotomy he underwent a few years ago. His first words to me, in a disembodied electronic tone were, "Welcome. Will you have coffee?"

As we drank—Hawking took tiny sips from a cup held up by one of his nurses—the interview fell into a pattern. First, I'd ask a question. Then I'd sit for ten or fifteen minutes while Hawking laboriously touch-screened his answer with his one mobile finger. Finally, the electronic voice would respond with at most a sentence or two.

I asked Hawking why he had called the COBE result the greatest discovery of the century, if not of all time. His answer was simply that COBE's measurements were consistent with (although by no means a proof of) inflation. His public statement at the time, which may have helped George Smoot get a reportedly huge advance for a book he subsequently cowrote about the discovery, evidently meant no more than that. It was perhaps a stronger statement about Hawking's belief in inflation than about the actual significance of COBE's findings.

Finally, the astronomers who had been skeptical about most of the cosmological models, and who were dubious about how close the final answer might be, continued to feel that way. This was especially true of Ed Turner at Princeton, whose pronouncement at the end of the Princeton post-COBE conference had seemed so reasonable. "I think the moral I can draw from what I've heard over the past few days," he said, "is that being close to the truth but wrong might mean we're close but it might mean we're wrong."

I managed to catch up with Turner between his latest conference and a family vacation that would end up (for him) with another observing run at Kitt Peak. We sat in his living room while his wife, Joyce, explained to the housesitters how to run various appliances. He was, as ever, the soul of reason.

"There have been various ideas for how to fix the standard model," he began. "One idea is tilting the spectrum of fluctuations. It's fairly plausible that the spectrum of fluctuations need not be exactly one [that is, that the degree of density fluctuation should be comparable at all angular scales]. One was a first approximation—but it's not unusual that when you get the data you go back and recalculate, and when people have done more precise calculations they've gotten the number lower, so that helps. Or maybe the cosmological constant is more significant than we think. Or . . . this is all the rage on the coast, as they say . . . you have mixed dark matter. Hot dark matter, cold dark matter, and baryons. And then there are still the heretical models—Peebles' model where omega is small and inflation is wrong, or textures.

"All of these are out on the table in a scramble. None has emerged as an obvious successor to the standard model. Most of the time, people still use the standard model. It's pretty clear it can't be exactly right, but the overall picture is not clear. My intuition is that none of them is exactly right. I don't sense much enthusiasm for any of these models."

His intuition was seconded by David Spergel a few days later. "People love their babies," Spergel said, "but . . . there is no model that's simple and fits the observations nicely. People at conferences keep apologizing for introducing fudge factors to make the models work, or for ignoring some observations. It's not clear whether we're very close or very far away"—essentially what Ed Turner had told the audience in Princeton two years before. When I called George Blumenthal, Joel Primack's Santa Cruz colleague and longtime collaborator, he said it explicitly:

"Has anything changed? Nothing radical. But," he added, "there are some smoke signals on the horizon."

The smoke signals, as usual, were in the form of new observations. For while the theorists strained mightily to incorporate the COBE data into existing models and to come up with new ones, the observers kept up their steady campaign to tease more information from the universe. David Wilkinson, George Smoot, Lyman Page, and the other microwave people continued to send up balloons and scan the South Polar skies and push NASA to build a new COBE, more powerful and capable of sampling the universe at much smaller angular scales; John Huchra and Bob Kirshner and their colleagues kept on with redshift surveys; the Princeton crew moved ahead with their own digital sky survey; Tony Tyson kept taking super-deep images of the night sky for subtle evidence of gravitational lenses; Sandra Faber kept trying to widen her mental map of the universe, and so on. At the same time, powerful new instruments like the Keck Telescope began taking data, while the unexpectedly successful Hubble Space Telescope repair mission restored that crippled instrument to a condition NASA Administrator Daniel Goldin called, with some justification, "better than new." As others had said before, and as Joel Primack reiterated, "If you believe all the data, then all the theories are ruled out. But not all the observations are correct." The only way to find out the truth is to make more and more careful observations.

In fact, the observers' steady industry has produced several new results for the theorists to chew on—some of them destined to be significant, others mistaken. Among them:

—The Harvard-Smithsonian redshift surveys were extended to the Southern Hemisphere, and they confirmed that the other half of the universe has walls and voids too.

—Two different teams searching for MACHOs—dim, massive, compact objects—in the halo of the Milky Way detected the characteristic signal of MACHOs—passing in front of stars in the Large Magellanic Cloud. There are probably too few of them, though, to account for the halo of dark matter that surrounds the galaxy, and some astronomers have argued that the objects aren't in our galaxy anyway, but are instead dim stars in the LMC itself.

—Princeton physicist Ruth Daly, one of the regulars at the Institute for Advanced Study's Tuesday Lunch, proposed a new test to see whether the universe is destined to expand forever or not. She measured the apparent size of radio-emitting jets in active galaxies, and calculated how their size changed with their distance from Earth. The size falls off with distance, but how quickly it does so depends on the curvature of space. By Daly's analysis the universe is open, and thus contains too little matter to halt the cosmic expansion. The result is far from generally accepted.

—Several astronomers took a picture of the spiral galaxy NGC 5907, which is about 35 million light-years from Earth, and found a faint glow of what looks like starlight that forms a halo around the edge-on disk; the halo seems to have about the same form and extent as the dark-matter halos that surround virtually all spirals. Whether the exceedingly feeble glow actually comes from stars, whether it can give astronomers any clues about dark matter, whether it is coming from the dark matter itself, or whether it is real at all are questions whose answers will have to wait for more observations.

—A refined analysis of gravitational-lens statistics by Chris Kochanek at Harvard, comparing the number of quasars that are lensed with the number that are are not, has come up with the same result that John Bahcall got with the prerepair Hubble Telescope: the cosmological constant, which might have made up some of the difference between the omega astronomers can measure and the figure of one that they would like to see, probably isn't big enough to do the job.

—Several observers began turning up increasing numbers of what are known as low-surface-brightness galaxies. They put out plenty of light, but the light is spread out evenly across their surfaces. The result is that these galaxies are exceedingly hard to see. Once astronomers knew what to look for, though, they began seeing low-surface-brightness galaxies all over the place. There may be enough of them to explain why there had seemed to be so many more of Tony Tyson's faint blue galaxies in the distant, early universe than there are galaxies in the modern universe. We may just not have counted carefully enough close to home.

—The COBE satellite itself kept taking data, and when the second year's observations were added to the first, one set of numbers changed dramatically. The quadrupole term, the measurement of temperature differences from one quarter of the sky to the next, dropped from about seventeen microkelvins to about six, while differences at smaller scales stayed about the same. Nobody is quite sure what this means, although the implications are not great for inflation, which had predicted no significant differences in the strength at any scale. "If you believe in the original predictions," David Spergel said, "then there's only about a 5% chance you'd see something like this. That's not so unlikely that you'd never expect to see it at all, but it's unlikely enough that it makes you nervous."

Is this last observation enough to prove that inflation never happened—enough to make Hawking eat his words about the greatest discovery of all time? After all, when Spergel had asked the assembled astrophysicists at the post-COBE Princeton conference what it would take to falsify inflation, Paul Steinhardt, one of the seminal thinkers on the topic, called out: "A totally incompatible spectrum of fluctuations." But this spectrum was not necessarily incompatible. "Inflation turns out to be a big tent," said Spergel. Its requirements, that is, turn out to be less strict than they first appear. Another way you could falsify inflation, according to several members of the audience, would be to discover that the universe is open rather than closed, that there really is only enough dark matter to account for the peculiar motions and rotations of galaxies but not enough to halt the expansion. Yet Spergel was now arguing that if you accept an open universe, then you can keep inflation and make the whole thing compatible with the newly analyzed COBE data. "I'm starting to think a lot about open-universe models myself," said Spergel. On the other hand, as Neil Turok had said at Princeton, with no contradiction from anyone, we don't really know that inflation happened at all.

There was one specter raised at the Princeton meeting, an observation that could throw all existing models—even after they'd been adjusted to account for COBE—into jeopardy. If you were to find bulk flows of galaxies, like the movements measured by the Seven Samurai, but on much larger scales, the whole Standard Model would be in real trouble once again. Just a few months after the COBE announcement and the subsequent furor, two observers announced precisely that kind of observation. Tod Lauer of the National Optical Astronomy Observatories and Marc Postman of the Space Telescope Science Institute had been quietly measuring the peculiar velocities of 119 galaxies, out to a distance three or four times as great as the Samurai had gone. At that scale, whatever the local motions caused by the gravity from unusual concentrations of matter, there should be essentially no motion with respect to the cosmic microwave background. But Postman and Lauer found, to their shock, that the entire collection of galaxies was moving toward a point about 50 degrees away from where the Great Attractor lies on the sky, at a velocity of about 700 kilometers per second, a little faster than the flow the Samurai had found.

"When we first got the result," Lauer told me over the telephone, "we were really kind of upset, it was so unexpected. We spent a year going over the data before we submitted the paper, trying to anticipate and answer any criticism anyone might have. So far, two years after our results appeared, we still haven't gotten any questions we can't answer. But some people just don't like to contemplate what we found, so they say, 'we can't see what you did wrong, but we still don't believe it.'

"In general, the California axis isn't too happy with us. The East Coast is a lot more receptive—people at Harvard and Princeton. I think the Princeton people, especially, have seen theories come and go over the years. They tend to be theorists there, but they're very willing to look at observations that contradict the conventional wisdom."

"It's a really pretty bit of work," agreed Ed Turner when I asked him about it, "one of the best in the last few years. It's also a distressing result. It really begins to be hard to understand how these things form." Out along the California axis, where the climate (for this result, at least) is presumably harsher, George Blumenthal didn't seem especially hostile.

"I think it's great," he said. "It's really impossible to reconcile with existing models. Most of the observations that are hard to explain are not that far off the predictions, but this is way, way off. The part of me that wants things to work is kind of worried, but the pixie inside of me who wants to upset everyone is really happy."

Blumenthal mentioned some other potentially disturbing observations. A rumor was going around the astronomical community about the Hubble Constant. As Ed Turner, Jackie Hewitt, and Tony Tyson tried to measure the distance to faraway galaxies with gravitational lenses and thus determine the expansion rate of the universe, and as Bob Kirshner and John Tonry tried to do the same by measuring the expanding fireballs of supernovas and by calculating the brightness of average stars in galaxies too far away for stars to be resolved, respectively, several groups were using the more traditional—and still the most accurate—method: measuring the apparent brightness of Cepheid variable stars.

With the Hubble Telescope fixed, it suddenly became possible to do that in much more distant galaxies, where the Hubble expansion overwhelms any local motions, and thereby to calculate the distance to these galaxies directly. "What I hear," said Blumenthal, "is that the number is going to come in high, about eighty [that is, 80 kilometers per second of additional speed per megaparsec of distance]." That is very bad news for people who believe omega is one. ("It's our big nightmare," Joel Primack had told me.) The reason is that a quickly expanding universe has to be relatively young. And if omega is one, then it has already decelerated, which means that the universe's average speed of expansion over its lifetime was even faster.

"If omega is one, and if the Hubble Constant is eighty," said Blumenthal, "then the universe is nine billion years old, and the stars in globular clusters are thirteen or fourteen billion. If anything, I've become more confident lately of the latter estimates, so that discomfort is growing."

The second of Blumenthal's smoke signals came from new calculations, made by several different groups, of the amount of deuterium and helium in the early universe. In brief, they had found evidence that the early universe had produced less ordinary matter than anyone had thought. That means that there has to be even more dark matter, cold or hot or both, to bring the total omega up to one.

Finally, there was a rumor that I heard from someone who did not want to be identified as the source of still very secret information. A laboratory had tried once again to measure the mass of the neutrino, and this time had found it. The mass would turn out to be one electron-volt, a value that would make Primack quite happy since it fit nicely into the cold-plus-dark-matter theory. There was no way, however, to confirm this information before its official release, so Primack and I and everyone else simply had to wait.

The new results will be thrown in with the old, and, like the old, some will turn out to be right and some wrong. Models will be readjusted, and probably some of these, too, will finally be abandoned by their most ardent supporters when the data squeeze them into untenability. At some point, presumably, the observations and the theories will finally converge, and a new standard model of the universe will emerge. But for now, the revolution is still in full swing and the outcome is still in doubt.

"You could ask," said Ed Turner, " 'what do we really need to answer the question of how structure arose? Is there some critical data missing? Or should we be looking for a new insight, a critical idea, to make sense of the data we have?'

"My hunch is the latter. We usually complain about a lack of data, but the amount of new data over the course of the past twenty years is really astounding—COBE, X-ray data, redshift surveys. Yet I feel we're still about the same distance from solving this as we were when I was a grad student.

"It reminds me of a story. Jackie Hewitt and I were in Japan, and we went to have a look at a radio telescope located in the Japanese mountains. We didn't know exactly where it was, just the town that was nearby. We got to the town unannounced. This was before I knew much Japanese. All we could say was 'telescope.' So we started wandering around, saying the word. It was kind of foggy, and we kept getting lost, finding ourselves in cabbage fields and that sort of thing.

"Finally we found it, and got one of the graduate students to show us around. At one point, Jackie asked, 'What is the limiting resource here?' That is to say, what is the weak link in doing science—are the computers not powerful enough to process the data, or are there too few scientists, or what? The student didn't understand at first, but gradually she realized what Jackie meant. 'Limiting resource,' she said, 'is thinking.' I think that's where we are right now. We have lots of data, but need some new ideas. Sometimes they just come along—inflation was like that. The problems it addresses were well known, and yet it wasn't designed to address most of them."

David Spergel agreed that there is simply no way of knowing whether astronomers are close to solving the mysteries of what the universe is made of and how it evolved. "You'll pick up a book from 1994 ten years from now, and either you'll say, 'Boy, they just about had it,' or you'll say, 'Wow, those guys were way off.' Everything is up in the air. I'll bet the observations over the next decade will be just as exciting as those from the past decade. It's a wonderful time to be doing cosmology."

Numbers in *italics* indicate photos